NEWS

NASA

NATIONAL AERONAUTICS AND SPACE ADMINISTRATION
WASHINGTON, D.C. 20546

TELS. WO 2-4155
WO 3-6925

FOR RELEASE: THURSDAY A.M.
July 15, 1971

RELEASE NÓ: 71-119K

PRESS KIT

PROJECT: APOLLO 15
(To be launched no
earlier than July 26)

I0030985

contents

TABLES AND ILLUSTRATIONS

Cover: Apollo 15 on the Launch Pad
Lightening flashes in the sky behind the Saturn V rocket that will propel Apollo 15 to the moon, July 25, 1971.

Image Credit: NASA

Published by Books Express Publishing
Copyright © Books Express, 2012
ISBN 978-1-78039-864-8

Books Express publications are available from all good retail and online booksellers. For publishing proposals and direct ordering please contact us at: info@books-express.com

NEWS

NASA

NATIONAL AERONAUTICS AND SPACE ADMINISTRATION   (202) 962-4155
WASHINGTON, D.C. 20546   TELS: (202) 963-6925

Ken Atchison/Howard Allaway   **FOR RELEASE:**   THURSDAY, A.M.
(Phone 202/962-0666)   July 15, 1971

RELEASE NO: 71-119

APOLLO 15 LAUNCH JULY 26

The 12-day Apollo 15 mission, scheduled for launch on July 26 to carry out the fourth United States manned exploration of the Moon, will:

- Double the time and extend tenfold the range of lunar surface exploration as compared with earlier missions;

- Deploy the third in a network of automatic scientific stations;

- Conduct a new group of experiments in lunar orbit; and

- Return to Earth a variety of lunar rock and soil samples.

Scientists expect the results will greatly increase man's knowledge both of the Moon's history and composition and of the evolution and dynamic interaction of the Sun-Earth system.

This is so because the dry, airless, lifeless Moon still bears records of solar radiation and the early years of solar system history that have been erased from Earth. Observations of current lunar events also may increase understanding of similar processes on Earth, such as earthquakes.

-more-   6/30/71

The Apollo 15 lunar module will make its descent over the Apennine peaks, one of the highest mountain ranges on the Moon, to land near the rim of the canyon-like Hadley Rille. From this Hadley-Apennine lunar base, between the mountain range and the rille, Commander David R. Scott and Lunar Module Pilot James B. Irwin will explore several kilometers from the lunar module, driving an electric-powered lunar roving vehicle for the first time on the Moon.

Scott and Irwin will leave the lunar module for three exploration periods to emplace scientific experiments on the lunar surface and to make detailed geologic investigations of formations in the Apennine foothills, along the Hadley Rille rim, and to other geologic structures.

The three previous manned landings were made by Apollo 11 at Tranquillity Base, Apollo 12 in the Ocean of Storms and Apollo 14 at Fra Mauro.

The Apollo 15 mission should greatly increase the scientific return when compared to earlier exploration missions. Extensive geological sampling and survey of the Hadley-Apennine region of the Moon will be enhanced by use of the lunar roving vehicle and by the improved life support systems of the lunar module and astronaut space suit. The load-carrying capacity of the lunar module has been increased to permit landing a greater payload on the lunar surface.

-more-

Additionally, significant scientific data on the Earth-Sun-Moon system and on the Moon itself will be gathered by a series of lunar orbital experiments carried aboard the Apollo command/service modules. Most of the orbital science tasks will be accomplished by Command Module Pilot Alfred M. Worden, while his comrades are on the lunar surface.

Worden is a USAF major, Scott a USAF colonel and Irwin a USAF lieutenant colonel.

During their first period of extravehicular activity (EVA) on the lunar surface, Scott and Irwin will drive the lunar roving vehicle to explore the Apennine front. After returning to the LM, they will set up the Apollo Lunar Surface Experiment Package (ALSEP) about 300 feet West of the LM.

Experiments in the Apollo 15 ALSEP are: passive seismic experiment for continuous measurement of moonquakes and meteorite impacts; lunar surface magnetometer for measuring the magnetic field at the lunar surface; solar wind spectrometer for measuring the energy and flux of solar protons and electrons reaching the Moon; suprathermal ion detector for measuring density of solar wind high and low-energy ions; cold cathode ion gauge for measuring variations in the thin lunar atmosphere; and the heat flow experiment to measure heat emanating from beneath the lunar surface.

Scott and Irwin will use for the first time a percussive drill for drilling holes in the Moon's crust for placement of the heat flow experiment sensors and for obtaining samples of the lunar crust.

Additionally, two experiments independent of the ALSEP will be set up near the LM. They are the solar wind composition experiment for determining the isotopic makeup of noble gases in the solar wind; and the laser ranging retro-reflector experiment which acts as a passive target for Earth-based lasers in measuring Earth-Moon distances over a long-term period. The solar wind composition experiment has been flown on all previous missions, and the laser reflector experiment was flown on Apollos 11 and 14. The Apollo 15 reflector has three times more reflective area than the two previous reflectors.

The second EVA will be spent in a lengthy geology traverse in which Scott and Irwin will collect documented samples and make geology investigations and photopanoramas at a series of stops along the Apennine front.

The third EVA will be a geological expedition along the Hadley Rille and northward from the LM.

At each stop in the traverses, the crew will re-aim a high-gain antenna on the lunar roving vehicle to permit a television picture of their activities to be beamed to Earth.

A suitcase-size device -- called the lunar communications relay unit -- for the first time will allow the crew to explore beyond the lunar horizon from the LM and still remain in contact with Earth. The communications unit relays two-way voice, biomedical telemetry and television signals from the lunar surface to Earth. Additionally, the unit permits Earth control of the television cameras during the lunar exploration.

Experiments in the Scientific Instrument Module (SIM) bay of the service module are: gamma-ray spectrometer and X-ray fluorescence which measure lunar surface chemical composition along the orbital ground track; alpha-particle spectrometer which measures alpha-particles from radioactive decay of radon gas isotopes emitted from the lunar surface; mass spectrometer which measures the composition and distribution of the lunar atmosphere; and a subsatellite carrying three experiments which is ejected into lunar orbit for relaying scientific information to Earth on the Earth's magnetosphere and its interaction with the Moon, the solar wind and the lunar gravity field.

The SIM bay also contains equipment for orbital photography including a 24-inch panoramic camera, three-inch mapping camera and a laser altimeter for accurately measuring spacecraft altitude for correlation with the mapping photos.

Worden will perform an inflight EVA to retrieve the exposed film. Selected flight experiments will be conducted during transearth coast.

Scheduled for launch at 9:34 a.m. EDT, July 26, from NASA's Kennedy Space Center, Fla., the Apollo 15 will land on the Moon on Friday July 30. The lunar module will remain on the surface about 67 hours. Splashdown will be at 26.1° North latitude by 158° West longitude in the North Central Pacific, north of Hawaii.

The prime recovery ship for Apollo 15 is the helicopter landing platform USS Okinawa.

Apollo 15 command module call sign is "Endeavour," and the lunar module is "Falcon." As in all earlier lunar landing missions, the crew will plant an American Flag on the lunar surface near the landing point. A plaque with the date of the Apollo 15 landing and signatures of the crew will be affixed to the LM front landing gear.

Apollo 15 backup crewmen are USN Capt. Richard F. Gordon, Jr., commander; Mr. Vance Brand, command module pilot; and Dr. Harrison H. Schmitt, lunar module pilot.

APOLLO 15
INCREASED OPERATIONAL CAPABILITIES

IMPROVEMENT	SYSTEM	CAPABILITY
MOBILITY	• LUNAR ROVER VEHICLE	INCREASED RANGE, CREW MOBILITY, TRAVERSE PAYLOAD CAPACITY AND EFFICIENCY OF SURFACE OPERATIONS
EVA DURATION	• LCRU/GCTA • A7LB SUIT • -7PLSS	IMPROVED LIFE SUPPORT SYSTEM INCREASES TOTAL EVA DURATION FROM 18 TO 40 MANHOURS
SURFACE DURATION	• LM	VEHICLE MODIFICATIONS PERMITTED NOMINAL LUNAR SURFACE STAY TIME ABOUT DOUBLE. (FROM 37 TO 67 HOURS)
ORBITAL SCIENCE	• CM/SM	ADDED SIM BAY AND EXPERIMENT CONTROLS TO PERMIT CONDUCTING ADDITIONAL ORBITAL SCIENCE
PAYLOAD CAPABILITY	• SATURN V LAUNCH VEHICLE	CAPABILITY INCREASED TO ACCOMODATE THE INCREASED WEIGHT OF THE PRIOR ITEMS

MISSION COMPARISON SUMMARY

	APOLLO 14	APOLLO 15
LAUNCH WINDOWS	1-3-3	2-2-3
LAUNCH WINDOW DURATION	3.5 HOURS	2.5 HOURS
LAUNCH AZIMUTH	72 – 96 DEGREES	80 – 100 DEGREES
EARTH PARKING ORBIT	100 NM	90 NM
SPACECRAFT PAYLOAD	102,095 POUNDS	107,500 POUNDS
TRANSLUNAR TRAJECTORY	TRANSFER MANEUVER	NO TRANSFER MANEUVER
LUNAR ORBIT INCLINATION	14 DEGREES	26 DEGREES
SCIENTIFIC INSTRUMENT MODULE	NO	LUNAR ORBIT & TRANSEARTH
LUNAR DESCENT TRAJECTORY	16 DEGREES	25 DEGREES
POST LUNAR LANDING	EVA-1	SEVA AND SLEEP
EVA's	2 (4:45 AND 4:30)	3 (7-7-6)
LUNAR SURFACE STAY TIME	33.5 HOURS	67 HOURS
SUBSATELLITE DEPLOYMENT	NO	REV 74
TRANSEARTH EVA	NO	ONE HOUR
EARTH LANDING	27 DEGREES SOUTH	26 DEGREES NORTH
MISSION DURATION	9 DAYS	12 DAYS, 7 HOURS

-end-

COUNTDOWN

The Apollo 15 launch countdown will be conducted by a government-industry team of about 500 working in two control centers at the Kennedy Space Center.

Overall space vehicle operations will be controlled from Firing Room No. 1 in the Complex 39 Launch Control Center. The spacecraft countdown will be run from an Acceptance Checkout Equipment (ACE) room in the Manned Spacecraft Operations Building.

More than five months of extensive checkout of the launch vehicle and spacecraft components are completed before the space vehicle is ready for the final countdown. The prime and backup crews participate in many of these tests including mission simulations, altitude runs, a flight readiness test and a countdown demonstration test.

The space vehicle rollout -- the three and one-half-mile trip from the Vehicle Assembly Building to the launch pad -- took place May 11.

Apollo 15 will be the ninth Saturn V launch from Pad A (seven manned). Apollo 10 was the only launch to date from Pad B, which will be used again in 1973 for the Skylab program.

The Apollo 15 precount activities will start at T-5 days. The early tasks include electrical connections and pyrotechnic installation in the space vehicle. Mechanical buildup of the spacecraft is completed, followed by servicing of the various gases and cryogenic propellants (liquid oxygen and liquid hydrogen) to the CSM and LM. Once this is accomplished, the spacecraft batteries are placed on board and the fuel cells are activated.

The final countdown begins at T-28 hours when the flight batteries are installed in the three stages and instrument unit of the launch vehicle.

At the T-9 hour mark, a built-in hold of nine hours and 34 minutes is planned to meet contingencies and provide a rest period for the launch crew. A one hour built-in hold is scheduled at T-3 hours 30 minutes.

Following are some of the highlights of the latter part of the count:

T-10 hours, 15 minutes   Start mobile service structure (MSS) move to park site

T-9 hours	Built-in hold for nine hours and 34 minutes. At end of hold, pad is cleared for LV propellant loading.
T-8 hours, 05 minutes	Launch vehicle propellant loading - Three stages (LOX in first stage, LOX and LH_2 in second and third stages). Continues thru T-3 hours 38 minutes.
T-4 hours, 15 minutes	Flight crew alerted.
T-4 hours, 00 minutes	Crew medical examination.
T-3 hours, 30 minutes	Crew breakfast.
T-3 hours, 30 minutes	One-hour built-in hold.
T-3 hours, 06 minutes	Crew departs Manned Spacecraft Operations Building for LC-39 via transfer van.
T-2 hours, 48 minutes	Crew arrival at LC-39.
T-2 hours, 40 minutes	Start flight crew ingress.
T-1 hours, 51 minutes	Space Vehicle Emergency Detection System (EDS) test (Scott participates along with launch team).
T-43 minutes	Retract Apollo access arm to stand-by position (12 degrees).
T-42 minutes	Arm launch escape system. Launch vehicle power transfer test, LM switch to internal power.
T-37 minutes	Final launch vehicle range safety checks (to 35 minutes).
T-30 minutes	Launch vehicle power transfer test, LM switch over to internal power.
T-20 minutes to T-10 minutes	Shutdown LM operational instrumentation.
T-15 minutes	Spacecraft to full internal power.
T-6 minutes	Space vehicle final status checks.

T-5 minutes, 30 seconds	Arm destruct system.
T-5 minutes	Apollo access arm fully retracted.
T-3 minutes, 6 seconds	Firing command (automatic sequence).
T-50 seconds	Launch vehicle transfer to internal power.
T-8.9 seconds	Ignition start.
T-2 seconds	All engines running.
T-0	Liftoff.

NOTE: Some changes in the countdown are possible as a result of experience gained in the countdown demonstration test which occurs about two weeks before launch.

Launch Windows

Launch date	Windows (EDT) Open	Close	Sun Elevation Angle
July 26, 1971	9:34 am	12:11 pm	12.0° *
July 27, 1971 (T+24)	9:37 am	12:14 pm	23.2°
Aug. 24, 1971 (T-0)	7:59 am	10:38 am	11.3°
Aug. 25, 1971 (T+24)	8:17 am	10:55 am	22.5°
Sept. 22, 1971 (T-24)	6:37 am	9:17 am	12.0°
Sept. 23, 1971 (T-0)	7:20 am	10:00 am	12.0°
Sept. 24, 1971 (T+24)	8:33 am	11:12 am	23.0°

* Only for launch azimuth of 80°

-more-

Ground Elapsed Time Update

It is planned to update, if necessary, the actual ground elapsed time (GET) during the mission to allow the major flight plan events to occur at the pre-planned GET regardless of either a late liftoff or trajectory dispersions that would otherwise have changed the event times.

For example, if the flight plan calls for descent orbit insertion (DOI) to occur at GET 82 hours, 40 minutes and the flight time to the Moon is two minutes longer than planned due to trajectory dispersions at translunar injection, the GET clock will be turned back two minutes during the translunar coast period so that DOI occurs at the pre-planned time rather than at 82 hours, 42 minutes. It follows that the other major mission events would then also be accomplished at the pre-planned times.

Updating the GET clock will accomplish in one adjustment what would otherwise require separate time adjustments for each event. By updating the GET clock, the astronauts and ground flight control personnel will be relieved of the burden of changing their checklists, flight plans, etc.

The planned times in the mission for updating GET will be kept to a minimum and will, generally, be limited to three updates. If required, they will occur at about 53, 97 and 150 hours into the mission. Both the actual GET and the update GET will be maintained in the MCC throughout the mission.

Launch and Mission Profile

The Saturn V launch vehicle (SA-510) will boost the Apollo 15 spacecraft from Launch Complex 39A at the Kennedy Space Center, Fla., at 9:34 a.m. EDT, July 26, 1971, on an azimuth of 80 degrees.

The first stage (S-1C) will lift the vehicle 38 nautical miles above the Earth. After separation the booster will fall into the Atlantic Ocean about 367 nautical miles downrange from Cape Kennedy, approximately nine minutes, 21 seconds after liftoff.

The second stage (S-II) will push the vehicle to an altitude of about 91 nautical miles. After separation, the S-II stage will follow a ballistic trajectory as it plunges into the Atlantic about 2,241 nautical miles downrange from Cape Kennedy about 19 minutes, 41 seconds into the mission.

The single engine of the third stage (S-IVB) will insert the vehicle into a 90-nautical-mile circular parking orbit before it is cut off for a coasting period. When reignited, the engine will inject the Apollo spacecraft into a translunar trajectory.

Launch Events

Time Hrs	Min	Sec	Event	Vehicle Wt (Pounds)	Altitude (Feet)	Velocity (Ft/Sec)	Range (Nau Mi)
00	00	00	First Motion	6,407,758	198	0	0
00	01	20	Maximum Dynamic Pressure	4,048,843	42,869	1,605	3
00	02	15.8	S-1C Center Engine Cutoff	2,388,283	155,162	5,573	26
00	02	38.7	S-1C Outboard Engines Cutoff	1,841,856	225,008	7,782	48
00	02	40.5	S-1C/S-II Separation	1,477,783	230,893	7,799	50
00	02	42.2	S-II Ignition	1,477,782	236,196	7,778	52
00	03	10.5	S-II Aft Interstage Jettison	1,406,067	320,265	8,116	86
00	03	16.2	Launch Escape Tower Jettison	1,383,533	335,636	8,210	93
00	07	38.8	S-II Center Engine Cutoff	651,648	584,545	17,362	594
00	09	9.4	S-II Outboard Engines Cutoff	476,526	576,526	21,551	876
00	09	10.4	S-II/S-IVB Separation	476,155	576,535	21,560	880
00	09	13.5	S-IVB Ignition	377,273	576,529	21,564	890
00	11	38.8	S-IVB First Cutoff	309,898	563,570	24,233	1,422
00	11	48.8	Parking Orbit Insertion (90 nm)	309,771	563,501	24,237	1,461

Mission Events

Events	GET hrs:min	Date/EDT	Velocity change feet/sec	Purpose and resultant orbit
Translunar injection (S-IVB engine ignition)	02:56	26/12:30 pm	10,036	Injection into translunar trajectory with 68 nm pericynthion
CSM separation, docking	03:20	26/12:54 pm	--	Mating of CSM and LM
Ejection from SLA	04:15	26/01:49 pm	1	Separates CSM-LM from S-IVB/SLA
S-IVB evasive maneuver	04:39	26/2:13 pm	10	Provides separation prior to S-IVB propellant dump and thruster maneuver to cause lunar impact
Residual Propellant Dump	05:00	26/2:34 pm		
APS Impact Burn (4 min.)	05:45	26/3:19 pm		
APS Correction Burn	09:30	26/7:04 pm		
Midcourse correction 1	TLI+9 hrs	26/9:29 pm	0*	*These midcourse corrections have a nominal velocity change of 0 fps, but will be cal-culated in real time to correct TLI dis-persions; trajectory within capability of docked DPS burn should SPS fail to ignite.
Midcourse correction 2	TLI+28 hrs	27/4:29 pm	0	
Midcourse correction 3	LOI-22 hrs	28/6:05 pm	0*	
Midcourse correction 4	LOI-5 hrs	29/11:05 am	0*	
SIM Door jettison	LOI-4.5 hrs	29/11:35 am	9	
Lunar orbit insertion	78:33	29/4:07 pm (Thurs.)	-2,998	Inserts Apollo 15 into 58 X 170 nm elliptical lunar orbit

Events	GET hrs:min	Date/EDT	Velocity change feet/sec	Purpose and resultant orbit
S-IVB impacts lunar surface	79:13	29/4:47 pm	--	Seismic event for Apollo 12 and 14 passive seismometers
Descent orbit insertion (DOI)	82:40	29/8:14 pm	-207	SPS burn places CSM/LM into 8 x 58 nm lunar orbit
CSM-LM undocking	100:14	30/1:48 pm	--	
CSM circularization	101.35	30/3:09 pm	70	Inserts CSM into 54 X 65 nm orbit (SPS burn)
LM Powered descent initiation	104:29	30/6:03 pm	6,698	Three-phase DPS burn to brake LM out of transfer orbit, vertical descent and touchdown on lunar surface
LM touchdown on lunar surface	104:41	30/6:15 pm (Friday)	--	Lunar exploration, deploy ALSEP, collect geological samples, photography

APOLLO 15
25° APPROACH TRAJECTORY

- SIGNIFICANT ENHANCEMENT OF TERRAIN CLEARANCE

- SIGNIFICANT ENHANCEMENT OF VISIBILITY AND FIDELITY OF LPD

- NO SIGNIFICANT INCREASE IN VERTICAL VELOCITY

- MODEST INCREASE IN ΔV FOR REDESIGNATIONS

MPAD 71-527 F

POWERED DESCENT PROFILE

SUMMARY					
EVENT	TFI, MIN:SEC	V₁, FPS	Ḣ, FPS	ĥ, FT	ΔV, FPS
POWERED DESCENT INITIATION	0:00	5562	-5	50,087	0
THROTTLE TO MAXIMUM THRUST	0:26	5534	-4	49,979	28
YAW TO VERTICAL	3:00	4111	-58	44,040	1468
LANDING RADAR ALTITUDE UPDATE	4:06	3444	-67	39,878	2159
LANDING RADAR VELOCITY UPDATE	5:34	2500	-85	33,623	3167
THROTTLE RECOVERY	7:24	1163	-80	22,950	4597
HIGH GATE	9:24	318	-162	7,029	5640
LOW GATE	10:42	66 (76)*	-23	694	6241
LANDING	12:02	-15 (0)*	-5	5	6698

*(HORIZONTAL VELOCITY RELATIVE TO SURFACE)

APPROACH PHASE COMPARISION

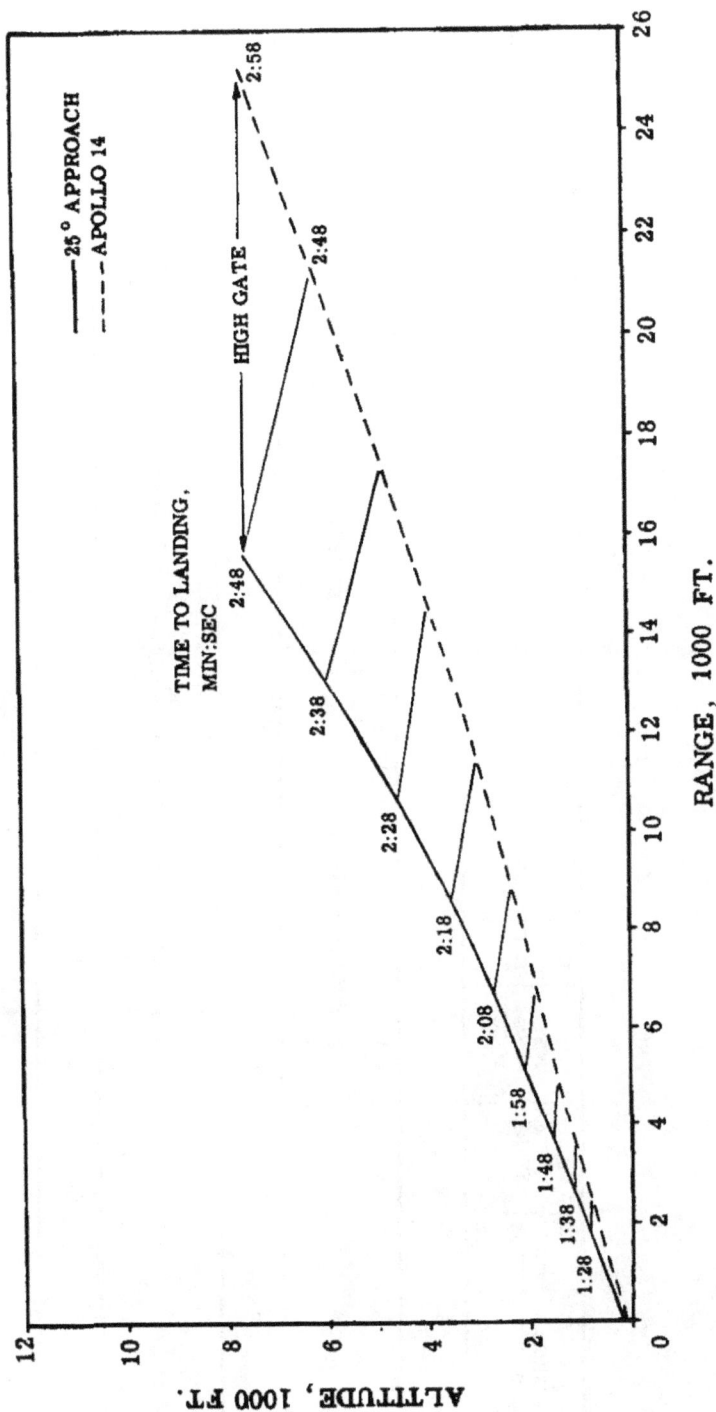

APOLLO 15 LUNAR SURFACE ACTIVITIES SUMMARY

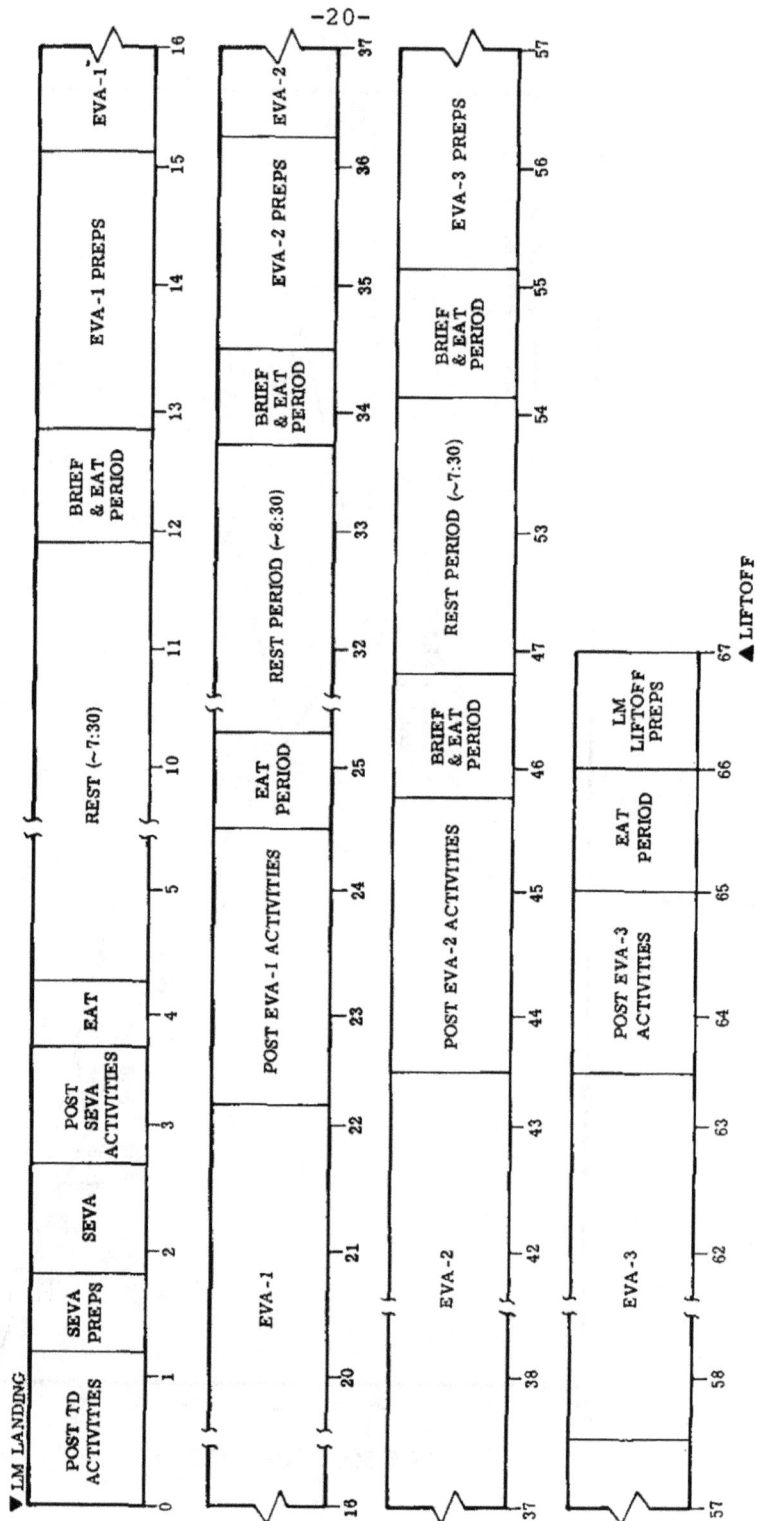

▼ LM LANDING

POST TD ACTIVITIES	SEVA PREPS	SEVA	POST SEVA ACTIVITIES	EAT	REST (~7:30)	BRIEF & EAT PERIOD	EVA-1 PREPS	EVA-1
0	1	2	3	4	5 ... 10 11 12	13	14 15	16

EVA-1 — 16 20 21 22 23 24 25

POST EVA-1 ACTIVITIES | EAT PERIOD | REST PERIOD (~8:30) | BRIEF & EAT PERIOD | EVA-2 PREPS | EVA-2

32 33 34 35 36 37

EVA-2 — 37 38 42 43 44 45 46 47

POST EVA-2 ACTIVITIES | BRIEF & EAT PERIOD | REST PERIOD (~7:30) | BRIEF & EAT PERIOD | EVA-3 PREPS

53 54 55 56 57

EVA-3 — 57 58 62 63 64 65 66 67

POST EVA-3 ACTIVITIES | EAT PERIOD | LM LIFTOFF PREPS

▲ LIFTOFF

EVA Mission Events

Events	GET hrs:min	Date/EDT
CDR starts standup EVA (SEVA) for verbal description of landing site, 360° photopanorama	106:10	Jul 30/7:44 pm
End SEVA, repressurize	106:40	8:14 pm
Depressurize LM for EVA 1	119:50	Jul 31/9:24 am
CDR steps onto surface	120:05	9:39 am
LMP steps onto surface	120:14	9:48 am
CDR places TV camera on tripod	120:16	9:50 am
LMP collects contingency sample	120:17	9:51 am
LMP climbs LM ladder to leave contingency sample on platform	120:20	9:54 am
Crew unstows LRV	120:20	9:54 am
LRV test driven	120:35	10:09 am
LRV equipment installation complete	120:58	10:32 am
Crew mounts LRV for drive to geology station No. 1--Hadley Rille rim near "elbow"; 2--base of Apennine front between "elbow" and St. George crater; 3--Apennine front possible debris flow area	121:12	10:46 am
Start LRV traverse back to LM	123:12	12:46 pm
Arrive at LM	123:40	1:14 pm
Offload ALSEP from LM, load drill and LRRR on LRV	123:58	1:32 pm

-more-

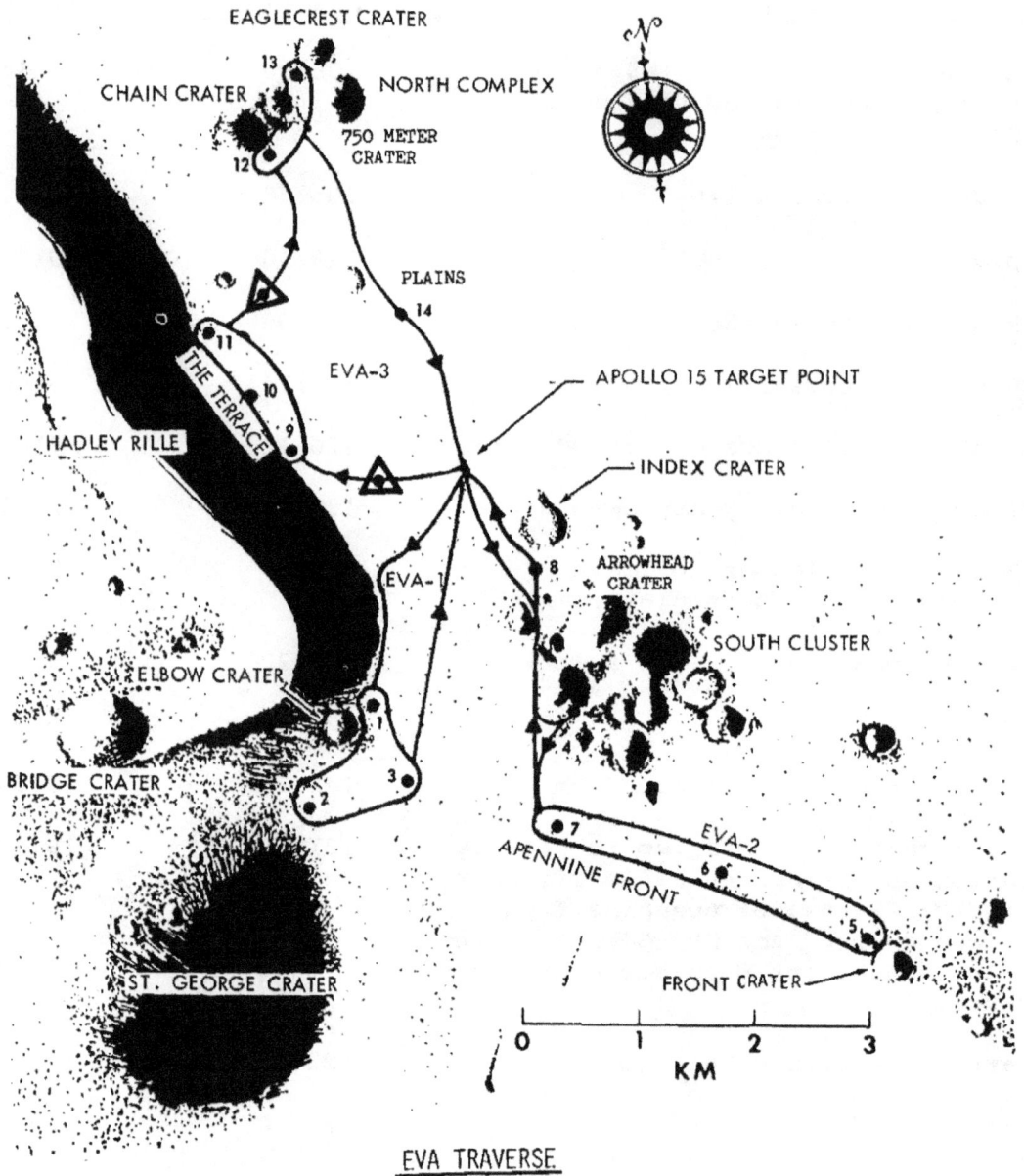

EAGLECREST CRATER

CHAIN CRATER

NORTH COMPLEX

750 METER CRATER

PLAINS

THE TERRACE

HADLEY RILLE

EVA-3

APOLLO 15 TARGET POINT

INDEX CRATER

ARROWHEAD CRATER

SOUTH CLUSTER

EVA-1

ELBOW CRATER

BRIDGE CRATER

ST. GEORGE CRATER

APENNINE FRONT

EVA-2

FRONT CRATER

0   1   2   3

KM

EVA TRAVERSE

TRAVERSE SUMMARY

TRAVERSE	START	END	DURATION	DISTANCE	RIDING TIME*	STATION TIME
I	1:25	3:50	2:25	7.9 KM	1:11	1:14
II	:49	6:20	5:31	16.1 KM	2:15	3:16
III	:42	5:15	4:33	12.3 KM	1:28	3:05
TOTALS				36.3 KM	4:54	7:35

• INCLUDES LRV INGRESS/EGRESS TIMES

TRAVERSE PLAN
EVA-1

STATION/AREA	ACTIVITY	STATION TIME
1 (ELBOW)	RADIAL SAMPLE	
2 (ST. ĠEORGE)	RADIAL SAMPLE COMPREHENSIVE SAMPLE 500mm PHOTOGRAPHY STEREO PAN PENETROMETER	
3	DOCUMENTED SAMPLE	
NEAR LM	ALSEP DEPLOYMENT LR^3 DEPLOYMENT SWC DEPLOYMENT MARE SAMPLING	

Events	GET hrs:min	Date/EDT
CDR drives LRV to ALSEP site, LMP walks	124:05	1:39 pm
Crew deploys ALSEP	124:08	1:42 pm
ALSEP deploy complete, return by LRV to LM	125:49	3:23 pm
Arrive at LM	125:55	3:29 pm
LMP deploys solar wind composition experiment, CDR makes polarimetric photos	125:58	3:32 pm
Crew erects US flag	126:13	3:47 pm
Crew stows equipment at LM and on LRV	126:18	3:52 pm
Crewmen dust lunar material from each other's EMUs	126:24	3:58 pm
LMP ingresses LM, CDR sends up Sample Return Container No. 1 on transfer conveyor	126:27	4:01 pm
CDR ingresses LM	126:40	4:14 pm
Repressurize LM, end EVA 1	126:50	4:24 pm
Depressurize LM for EVA 2	141:12	Aug 1/6:46 am
CDR steps onto surface	141:23	6:57 am
LMP steps onto surface	141:37	7:11 am
Crew loads gear aboard LRV for geology traverse, begin drive to Apennine front	141:59	7:33 am
Arrive secondary crater cluster (sta.4)	142:27	8:01 am
Arrive at Front Crater, gather samples, photos of front materials on crater rim	143:16	8:50 am

-more-

TRAVERSE PLAN
EVA-2

STATION/AREA	ACTIVITY	STATION TIME
4 (SECONDARIES)	SOIL/RAKE SAMPLE DOCUMENTED SAMPLE 500mm PHOTOGRAPHY EXPLORATORY TRENCH CORE TUBE (1)	
5 - 6 - 7 (SECONDARIES)	STATION 5: DOCUMENTED SAMPLES FROM UPSLOPE SIDE DOCUMENTED SAMPLES DOWNSLOPE SIDE EXPLORATORY TRENCH 500mm PHOTOGRAPHY STATION 6 - 7: DOCUMENTED SAMPLES EXPLORATORY TRENCHES CORE TUBE SAMPLE 500mm PHOTOGRAPHY	
8 (MARE)	COMPREHENSIVE SAMPLE DOUBLE CORE TUBE SAMPLE DOCUMENTED SAMPLE SESC TRENCH SOIL MECHANICS EXPERIMENT	

-more-

Events	GET hrs:min	Date/EDT
Arrive at area stop 5-6 on crater rim slope, samples, photos, soil mechanics trench	144:23	9:57 am
Arrive at stop 7--secondary crater cluster near 400m crater; collect documented samples, photopanorama	146:11	11:45 am
Arrive at stop 8 for investigations of materials in large mare area	146:47	12:21 pm
Arrive back at LM, hoist Sample Return Container No. 2 into LM	147:10	12:44 pm
Crew ingresses LM, repressurize, End EVA 2	148:10	1:44 pm
Depressurize for EVA 3	161:50	Aug 2/3:24 am
CDR steps onto surface	162:03	3:37 am
LMP steps onto surface	162:09	3:43 am
Prepare and load LRV for geology traverse	162:11	3:45 am
Leave for stations 9-13	162:44	4:18 am
Arrive station 9--rim of Hadley Rille; photos, penetrometer, core samples, documented samples	163:08	4:42 am
Arrive at station 10; documented samples, photopanorama	164:01	5:35 am
Arrive at station 11--rim of Hadley Rille; documented samples, photopanorama, description of near and far rille walls	164:17	5:51 am

-more-

TRAVERSE PLAN
EVA-3

STATION/AREA	ACTIVITY	STATION TIME
9 - 10 (RILLE)	STATION 9: 500mm PHOTOGRAPHY COMPREHENSIVE SAMPLE DOUBLE CORE DOCUMENTED SAMPLE SESC PENETROMETER	
	STATION 10: 500mm PHOTOGRAPHY DOCUMENTED SAMPLE	
11 (RILLE)	500mm PHOTOGRAPHY DOCUMENTED SAMPLE	
12 (N. COMPLEX/ CHAIN CRATER)	DOCUMENTED SAMPLE CORE TUBE	
13 (N. COMPLEX)	CRATER - DOCUMENTED SAMPLE - PHOTOGRAPHY SAMPLES, OBSERVATION & PHOTOGRAPHY OF: EAGLE CREST NORTH COMPLEX SCARPS	
14 (MARE)	DOCUMENTED SAMPLE	

Events	GET hrs:min	Date/EDT
Arrive at station 12--SE rim of Chain Crater; documented samples, photopanorama, seek unusual samples	165:00	6:34 am
Arrive at station 13--north complex scarp between larger craters; documented samples, photograph scarp, observe and describe 750m and 390m craters, core tubes, trench, penetrometer	165:31	7:05 am
Arrive station 14--fresh blocky crater in mare south of north complex; photopanorama, documented samples	166:43	8:17 am
Arrive back at LM	167:17	8:51 am
Load samples, film in LM; park LRV 300 feet east of LM, switch to ground-controlled TV for ascent	167:35	9:09 am
Crew ingress LM, end 3rd EVA	167:50	9:24 am

-more-

Mission Events (Cont'd.)

Events	GET hrs:min	Date/EDT	Velocity change feet/sec	Purpose and resultant orbit
CSM plane change	165:13	2/6:47 am	309	Changes CSM orbital plane by 3.3° to coincide with LM orbital plane at time of ascent from surface
LM ascent	171:35	2/01:09 pm	6,056	Boosts ascent stage into 9 X 46 nm lunar orbit for rendezvous with CSM
Insertion into lunar orbit	171:43	2/01:17 pm		
Terminal phase initiate (TPI) (LM APS)	172:30	2/2:04 pm	52	Boosts ascent stage into 61 X 44 nm catch-up orbit; LM trails CSM by 32 nm and 15 nm below at time of TPI burn
Braking (LM RCS; 4 burns)	173:11	2/2:45 pm	31	Line-of-sight terminal phase braking to place LM in 59 X 59 nm orbit for final approach, docking
Docking	173:30	2/3:04 pm		CDR and LMP transfer back to CSM
LM jettison, separation	177:38	2/7:12 pm		Prevents recontact of CSM with LM ascent stage during remainder of lunar orbit
LM ascent stage de-orbit (RCS)	179:06	2/8:40 pm	-195	ALSEP seismometers at Apollo 15, 14 and 12 landing sites record impact event
LM impact	179:31	2/9:05 am		Impact at about 5,528 fps at -4° angle, 32 nm from Apollo 15 ALSEP
CSM orbital change	221:25	4/2:59 pm	64	55 X 75 nm orbit (Rev 73)
Subsatellite ejection	222:36	4/4:10 pm		Lunar orbital science experiment

Events	GET hrs:min	Date/EDT	Velocity change feet/sec	Purpose and resultant orbit
Transearth injection (TEI) SPS	223:44	4/5:18 pm	3,047	Inject CSM into transearth trajectory
Midcourse correction 5	TEI+17 hrs	5/10:20 am	0	Transearth midcourse corrections will be computed in real time for entry corridor control and recovery area weather avoidance
Inflight EVA	242:00	5/11:34 am		To retrieve film cannisters from SM SIM bay
Midcourse correction 6	EI-22 hrs	6/6:32 pm	0	
Midcourse correction 7	EI-3 hrs	7/01:32 pm	0	
CM/SM separation	294:43	7/4:17 pm		Command module oriented for Earth atmosphere entry
Entry interface (400,000 ft)	294:58	7/4:32 pm		Command module enters atmosphere at 36,097 fps
Splashdown	295:12	7/4:46 pm		Landing 1,190 nm downrange from entry; splash at 26.1° North latitude, 158° West longitude

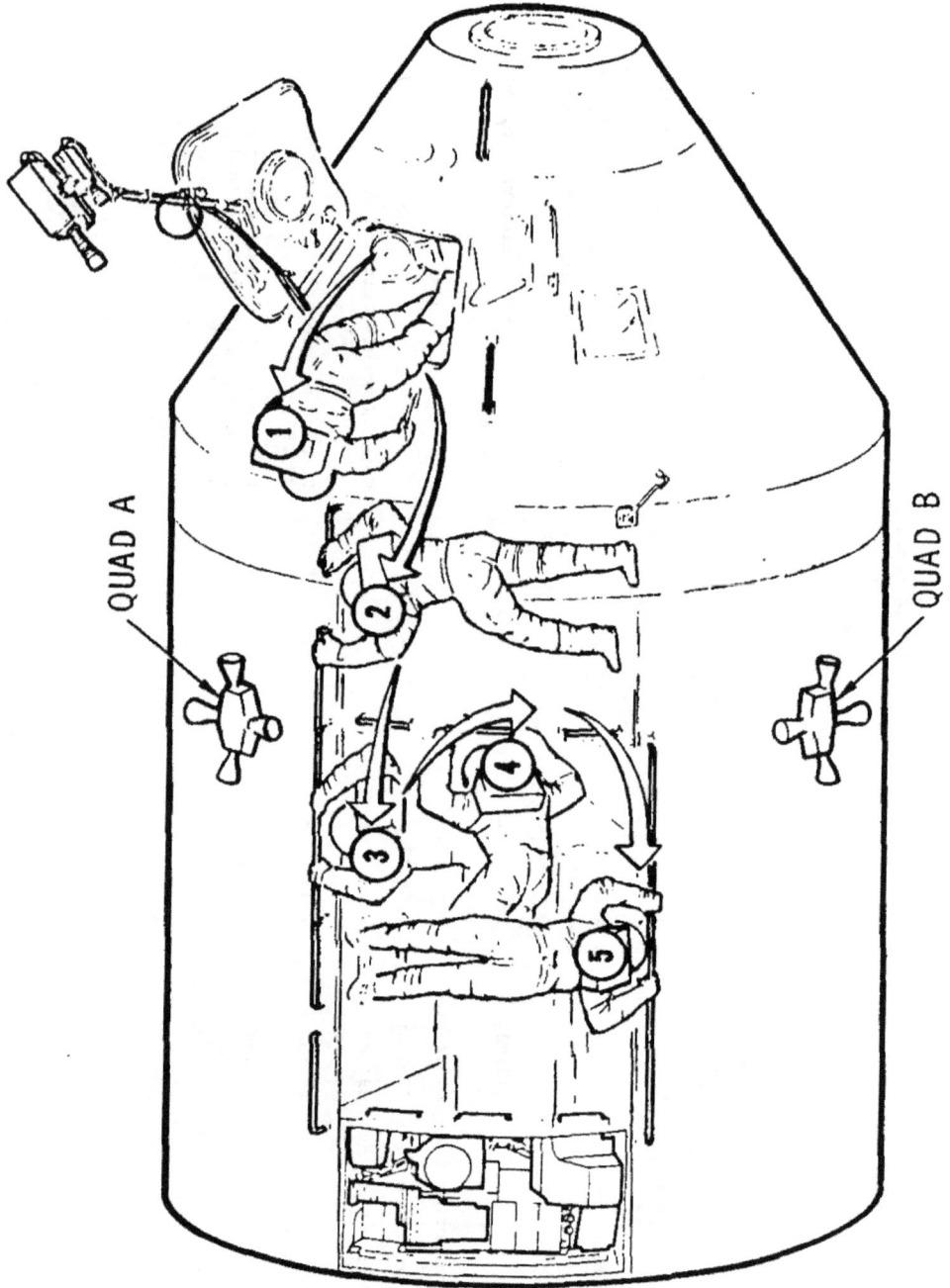

EVA PROCEDURES
CREWMAN PATH TO FOOT RESTRAINTS

QUAD A

QUAD B

Entry Events

Event	Time from 400,000 ft. min:sec	
Entry	00:00	4:32 p.m. 7th August
Enter S-band communication blackout	00:18	
Initiate constant drag	00:54	
Maximum heating rate	01:10	
Maximum load factor (FIRST)	01:24	
Exit S-band communication blackout	03:34	
Maximum load factor (SECOND)	05:42	
Termination of CMC guidance	06:50	
Drogue parachute deployment	07:47	(altitude, 23,000 ft.)
Main parachute deployment	08:36	(altitude, 10,000 ft.)
Landing	13:26	4:45 p.m. 7th August

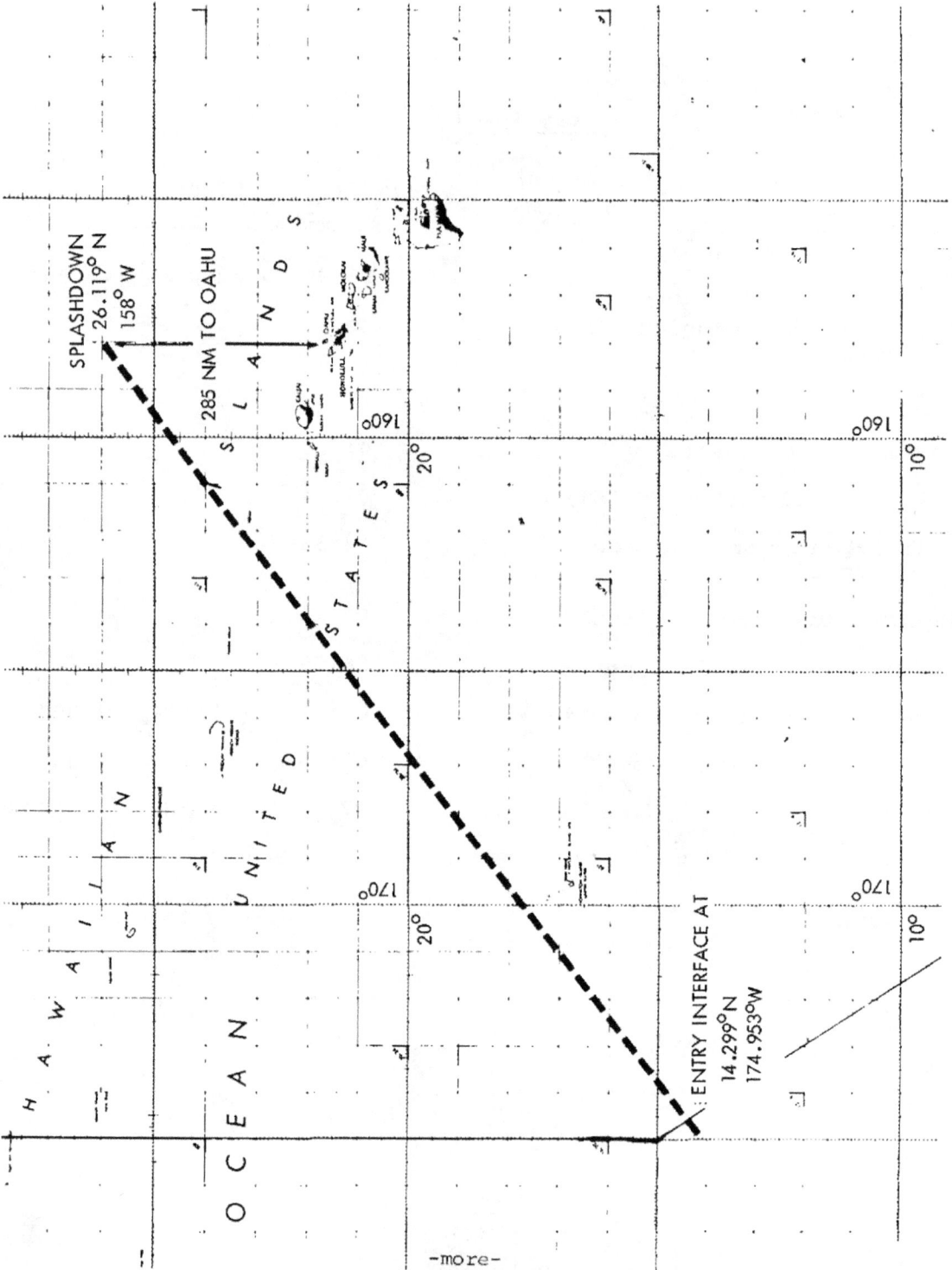

APOLLO 15 RECOVERY

SPLASHDOWN
26.119° N
158° W

285 NM TO OAHU

ENTRY INTERFACE AT
14.299° N
174.953° W

Recovery Operations

Launch abort landing areas extend downrange 3,400 nautical miles from Kennedy Space Center, fanwise 50 nm above and below the limits of the variable launch azimuth (80-100 degrees) in the Atlantic Ocean.

Splashdown for a full-duration lunar landing mission launched on time July 26 will be at 4:46 p.m. EDT, August 7 at 26.1° North latitude by 158° West longitude -- about 290 nm due north of Pearl Harbor, Hawaii.

The landing platform-helicopter (LPH) USS Okinawa, Apollo 15 prime recovery vessel, will be stationed near the end-of-mission aiming point prior to entry.

In addition to the primary recovery vessel located in the recovery area, HC-130 air rescue aircraft will be on standby at staging bases at Guam, Hawaii, Azores and Florida.

Apollo 15 recovery operations will be directed from the Recovery Operations Control Room in the Mission Control Center, supported by the Atlantic Recovery Control Center, Norfolk, Va., and the Pacific Recovery Control Center, Kunia, Hawaii.

The Apollo 15 crew will remain aboard the USS Okinawa until the ship reaches Pearl Harbor the day after splashdown. They will be flown from Hickam Air Force Base to Houston aboard a USAF transport aircraft. There will be no postflight quarantine of crew or spacecraft.

APOLLO 15 CREW POST-LANDING ACTIVITIES

DAYS FROM RECOVERY	DATE	ACTIVITY
SPLASHDOWN	AUGUST 7	
R+1	AUGUST 8	ARRIVE PEARL HARBOR
R+2	AUGUST 9	ARRIVE MSC
R+3 THRU R+15		CREW DEBRIEFING PERIOD
R + 5.	AUGUST 12	CREW PRESS CONFERENCE

APOLLO 15 MISSION OBJECTIVES

First of the Apollo J mission series which are capable of longer stay times on the Moon and greater surface mobility, Apollo 15 has four primary objectives which fall into the general categories of lunar surface science, lunar orbital science, and engineering/operational.

The mission objectives are to explore the Hadley-Apennine region, set up and activate lunar surface scientific experiments, make engineering evaluations of new Apollo equipment, and conduct lunar orbital experiments and photographic tasks.

Exploration and geological investigations at the Hadley-Appenine site will be enhanced by the addition of the lunar rover vehicle that will allow Scott and Irwin to travel greater distances from the lunar module than they could on foot during their three EVAs. Setup of the Apollo lunar surface experiment package (ALSEP) will be the third in a trio of operating ALSEPs (Apollos 12, 14, and 15.)

Orbital science experiments are primarily concentrated in an array of instruments and cameras in the scientific instrument module (SIM) bay of the spacecraft service module. Command module pilot Worden will operate these instruments during the period he is flying the command module solo and again for two days following the return of astronauts Scott and Irwin from the lunar surface. After transearth injection, he will go EVA to retrieve film cassettes from the SIM bay. In addition to operating SIM bay experiments, Worden will conduct other experiments such as gegenschein and ultraviolet photography tasks from lunar orbit.

Among the engineering/operational tasks to be carried out by the Apollo 15 crew is the evaluation of the modifications to the lunar module which were made for carrying a heavier payload and for a lunar stay time of almost three days. Changes to the Apollo spacesuit and to the portable life support system (PLSS) will be evaluated. Performance of the lunar rover vehicle (LRV) and the other new J-mission equipment that goes with it--the lunar communications relay unit (LCRU) and the ground-controlled television assembly (GCTA)--also will be evaluated.

LUNAR SURFACE EXPERIMENTS

EXPERIMENT	11	12	14	15
S-031 LUNAR PASSIVE SEISMOLOGY	X	X	X	X
S-033 LUNAR ACTIVE SEISMOLOGY			X	
S-034 LUNAR TRI-AXIS MAGNETOMETER		X		X
S-035 MEDIUM ENERGY SOLAR WIND		X		X
S-036 SUPRATHERMAL ION DETECTOR		X	X	X
S-037 LUNAR HEAT FLOW				X
S-038 CHARGED PARTICLE LUNAR ENVIRONMENT			X	
S-058 COLD CATHODE GAUGE		X	X	X
M-515 LUNAR DUST DETECTOR		X	X	X
S-059 LUNAR GEOLOGY INVESTIGATION	X	X	X	X
S-078 LASER RANGING RETRO-REFLECTOR	X		X	X
S-080 SOLAR WIND COMPOSITION	X	X	X	X
LUNAR SURFACE CLOSE-UP CAMERA	X	X	X	
S-198 LUNAR PORTABLE MAGNETOMETER			X	
S-200 SOIL MECHANICS				X

LUNAR ORBITAL EXPERIMENTS

		11	12	14	15
SERVICE MODULE					
S-160	GAMMA-RAY SPECTROMETER				X
S-161	X-RAY FLOURESCENCE				X
S-162	ALPHA-PARTICLE SPECTROMETER				X
S-164	S-BAND TRANSPONDER			X	X
S-165	MASS SPECTROMETER				X
S-170	BISTATIC RADAR			X	X
S-173	PARTICLE MEASUREMENT (SUBSATELLITE)				X
S-174	MAGNETOMETER (SUBSATELLITE)				X
S-164	S-BAND TRANSPONDER (SUBSATELLITE)				X
	24" PANORAMIC CAMERA				X
	3" MAPPING CAMERA				X
	LASER ALTIMETER				X
COMMAND MODULE					
S-176	APOLLO WINDOW METEOROID			X	X
S-177	UV PHOTOGRAPHY - EARTH AND MOON			X	X
S-178	GEGENSCHEIN FROM LUNAR ORBIT				X

Lunar Surface Science

As in previous lunar landing missions, a contingency sample of lunar surface material will be the first scientific objective performed during the first EVA period. The Apollo 15 landing crew will devote a large portion of the first EVA to deploying experiments in the ALSEP. These instruments will remain on the Moon to transmit scientific data through the Manned Space Flight Network on long-term physical and environmental properties of the Moon. These data can be correlated with known Earth data for further knowledge on the origins of the planet and its satellite.

The ALSEP array carried on Apollo 15 has seven experiments: S-031 Passive Seismic Experiment, S-034 Lunar Surface Magnetometer Experiment, S-035 Solar Wind Spectrometer Experiment, S-036 Suprathermal Ion Detector Experiment, S-037 Heat Flow Experiment, S-058 Cold Cathode Gauge Experiment, and M-515 Lunar Dust Detector Experiment.

Two additional experiments, not part of ALSEP, will be deployed in the ALSEP area: S-078 Laser Ranging Retro-Reflector and S-080 Solar Wind Composition.

Passive Seismic Experiment: (PSE): The PSE measures seismic activity of the Moon and gathers and relays to Earth information relating to physical properties of the lunar crust and interior. The PSE reports seismic data on man-made impacts (LM ascent stage), natural impacts of meteorites, and moonquakes. Dr. Gary Latham of the Lamont-Doherty Geological Observatory (Columbia University) is responsible for PSE design and experiment data analysis.

Two similar PSEs deployed as a part of the Apollo 12 and 14 ALSEPs have transmitted to Earth data on lunar surface seismic events since deployment. The Apollo 12, 14, and 15 seismometers differ from the seismometer left at Tranquillity Base in July 1969 by the Apollo 11 crew in that the later PSEs are continuously powered by SNAP-27 radioisotope electric generators. The Apollo 11 seismometer, powered by solar cells, transmitted data only during the lunar day, and is no longer functioning.

After Apollo 15 translunar injection, an attempt will be made to impact the spent S-IVB stage and the instrument unit into the Moon. This will stimulate the passive seismometers left on the lunar surface by other Apollo crews.

Background Scientific Information on the Lunar Surface Experiments

SCIENTIFIC DISCIPLINE / EXPERIMENT	GEOLOGY	GEOPHYSICS	GEOCHEMISTRY	BIOSCIENCES	GEODESY/CARTOGRAPHY	LUNAR ATMOSPHERE	PARTICLES AND FIELDS	ASTRONOMY
CONTINGENCY SAMPLE COLLECTION	AID IN DETERMINING LUNAR HISTORY BY AGING OF LUNAR SAMPLES		DETERMINE COMPOSITION OF LUNAR SURFACE BY CHEMICAL ANALYSIS OF LUNAR SAMPLES	AID IN DETERMINING POSSIBILITY OF BIOLOGICAL FORMS ON LUNAR SURFACE				
ALSEP PASSIVE SEISMIC (S-031)	AID IN DETERMINING INTERIOR STRUCTURE, TECTONISM AND VOLCANISM	AID IN DETERMINING TREE OSCILLATIONS, TIDES, SECULAR STRAINS, TILT, VELOCITY, GRAVITY CHANGES AND FREQUENCY AMPLITUDE, ATTENUATION AND DIRECTION OF SEISMIC WAVES						MEASURE METEOROID IMPACTS
HEAT FLOW (S-037)	AID IN DETERMINING LUNAR EVOLUTION FROM DATA	MEASURE VERTICAL TEMPERATURE GRADIENTS, ABSOLUTE TEMPERATURE OF THE SURFACE TO ESTABLISH VERTICAL THERMAL CONDUCTIVITY	DETERMINE BULK COMPOSITION AND CHEMICAL SORTING MAY BE INFERRED FROM DATA					DETERMINE THERMAL ENVIRONMENT
LUNAR SURFACE MAGNETOMETER (S-034)	AID IN DETERMINING MAGNETIC ANOMALIES, SUBSURFACE FEATURES AND LUNAR HISTORY	AID IN DETERMINING THERMAL STATE OF THE LUNAR INTERIOR					ESTABLISH GROSS ELECTRICAL DIFFUSIVITY, MEASURE MAGNETIC FIELD OF THE MOON	DETERMINE LUNAR RESPONSE TO FLUCTUATIONS IN THE INTERPLANETARY MAGNETIC FIELD
SOLAR WIND SPECTROMETER (S-035)		MONITOR FLUX, ENERGY STREAMING DIRECTION, AND TEMPORAL VARIATIONS IN THE SOLAR WIND PLASMA				DETERMINE PRESENCE OF ATMOSPHERE	ESTABLISH GROSS ELECTRICAL CONDUCTIVITY	
SUPRATHERMAL ION DETECTOR (S-036)		MEASURE FLUX, NUMBER DENSITY, VELOCITY AND ENERGY PER UNIT CHARGE OF POSITIVE IONS				DETERMINE IONOSPHERE/ATMOSPHERE CHARACTERISTICS	DETERMINE AMBIENT ELECTRIC FIELD EFFECTS	
COLD CATHODE ION GAUGE (S-058)						DETERMINE DENSITY OF THE LUNAR ATMOSPHERE INCLUDING TEMPORAL VARIATIONS	PROVIDE INFORMATION ON RATE OF LOSS ON CONTAMINANTS LEFT BY ASTRONAUTS	
LUNAR DUST DETECTOR (M-515)		AID IN DETERMINING SURFACE MATERIAL TRANSPORT, PROVIDE INFORMATION ON HIGH ENERGY RADIATION, DUST ACCUMULATION AND LUNAR SURFACE TEMPERATURES				PROVIDE INFORMATION ON DUST ACCUMULATION		
LASER RANGING RETRO-REFLECTOR (S-078)		DETERMINE FACTORS ABOUT LUNAR MOTION, LUNAR LIBRATION, AND EARTH GEOPHYSICAL DATA			AID IN DETERMINING EPHEMERIS, ORIENTATION AND LIBRATION			PROVIDE INCREASED ACCURACY IN LUNAR ORBITAL DATA PARAMETERS
LUNAR GEOLOGY INVESTIGATION (S-059)	AID IN DETERMINING LUNAR GEOLOGICAL STRUCTURE AND HISTORY		DETERMINE CHEMICAL COMPOSITION OF LUNAR SAMPLES	LUNAR SAMPLES MAY BE TESTED FOR ABILITY TO SUPPORT LIFE FORMS USED TO DETERMINE POSSIBILITY OF BIOLOGICAL LIFE FORMS ON THE LUNAR SURFACE				
SOIL MECHANICS (S-200)	AID IN DETERMINING LUNAR HISTORY, ENABLE DETERMINATION OF COMPOSITIONAL TEXTURAL AND MECHANICAL PROPERTIES OF LUNAR SOIL		ENABLE DETERMINATION OF COMPOSITION OF LUNAR SOIL					
SOLAR WIND COMPOSITION (S-080)		DETERMINE COMPOSITION OF SOLAR WIND PLASMA	DETERMINE COMPOSITION OF SOLAR WIND			AID IN DETERMINING HISTORY OF PLANETARY ATMOSPHERE	PROVIDE INFORMATION ON THE ELEMENTAL AND ISOTOPIC COMPOSITION OF NOBLE GASES AND OTHER ELEMENTS IN THE SOLAR WIND	

ALSEP ARRAY LAYOUT

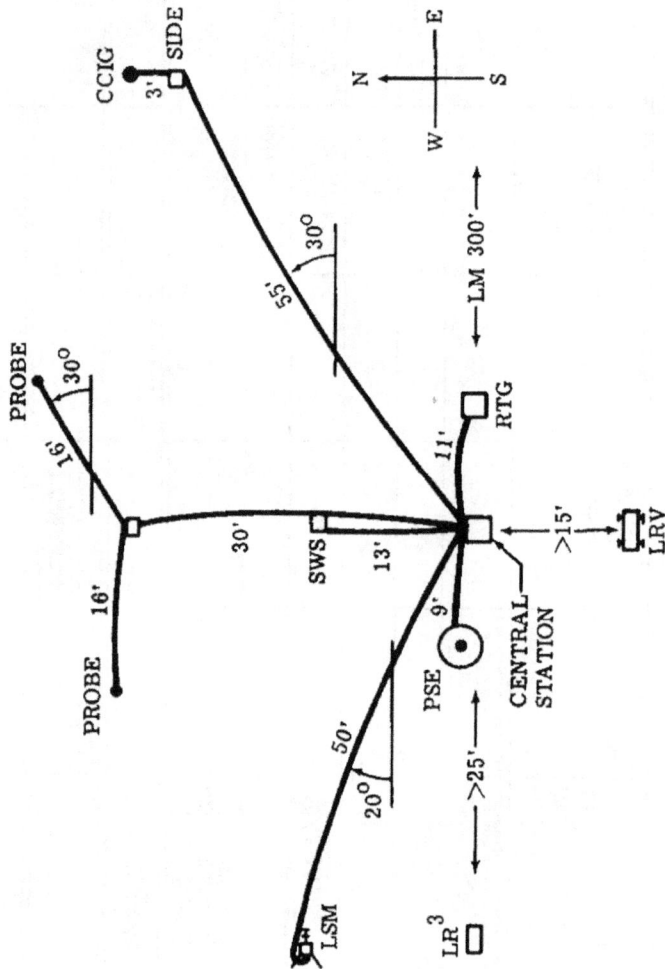

Through a series of switch-selection-command and ground-commanded thrust operations, the S-IVB/IU will be directed to hit the Moon within a target area 379 nautical miles in diameter. The target point is 3.65 degrees south latitude by 7.58 degrees west longitude, near Lelande Crater about 161 nautical miles east of Apollo 14 landing site.

After the lunar module is ejected from the S-IVB, the launch vehicle will fire an auxiliary propulsion system (APS) ullage motor to separate the vehicle from the spacecraft a safe distance. Residual liquid oxygen in the almost spent S-IVB/IU will then be dumped through the engine with the vehicle positioned so the dump will slow it into an impact trajectory. Mid-course corrections will be made with the stage's APS ullage motors if necessary.

The S-IVB/IU will weigh 30,836 pounds and will be traveling 4,942 nautical-miles-an-hour at impact. It will provide an energy source at impact equivalent to about 11 tons of TNT.

After Scott and Irwin have completed their lunar surface operations and rendezvoused with the command module in lunar orbit, the lunar module ascent stage will be jettisoned and later ground-commanded to impact on the lunar surface about 25 nautical miles west of the Apollo 15 landing site at Hadley-Apennine.

Impacts of these objects of known masses and velocities will assist in calibrating the Apollo 14 PSE readouts as well as providing comparative readings between the Apollo 12 and 14 seismometers forming the first two stations of a lunar surface seismic network.

There are three major physical components of the PSE:

1. The sensor assembly consists of three long-period and one short-period vertical seismometers with orthagonally-oriented capacitance-type seismic sensors, capable of measuring along two horizontal components and one vertical component. The sensor assembly is mounted on a gimbal platform. A magnet-type sensor short-period seismometer is located on the base of the sensor assembly.

2. The leveling stool allows manual leveling of the sensor assembly by the crewman to within ± 5 degrees. Final leveling to within ± 3 arc seconds is accomplished by control motors.

ALSEP to Impact Distance Table

Approximate Distance in:	Km	Statute Miles
Apollo 12 ALSEP to:		
Apollo 12 LM A/S Impact	75	45
Apollo 13 S-IVB Impact	134	85
Apollo 14 S-IVB Impact	173	105
Apollo 14 LM A/S Impact	116	70
Apollo 15 S-IVB Impact	480	300
Apollo 15 LM A/S Impact	1150	710
Apollo 14 ALSEP to:		
Apollo 14 LM A/S Impact	67	40
Apollo 15 S-IVB Impact	300	185
Apollo 15 LM A/S Impact	1070	660
Apollo 15 ALSEP to:		
Apollo 15 LM A/S Impact	50	30

S-IVB/IU IMPACT

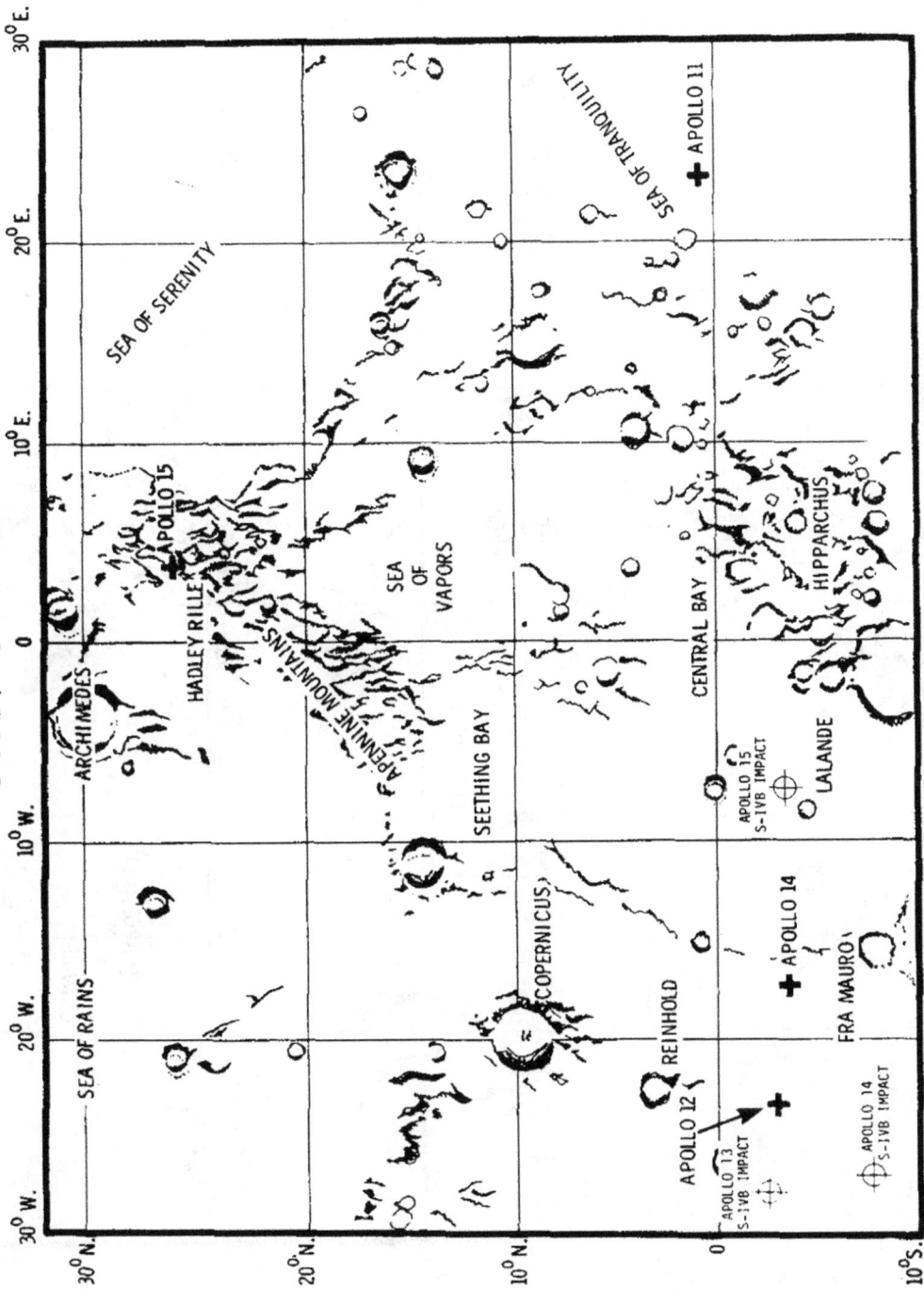

LM ASCENT STAGE IMPACT

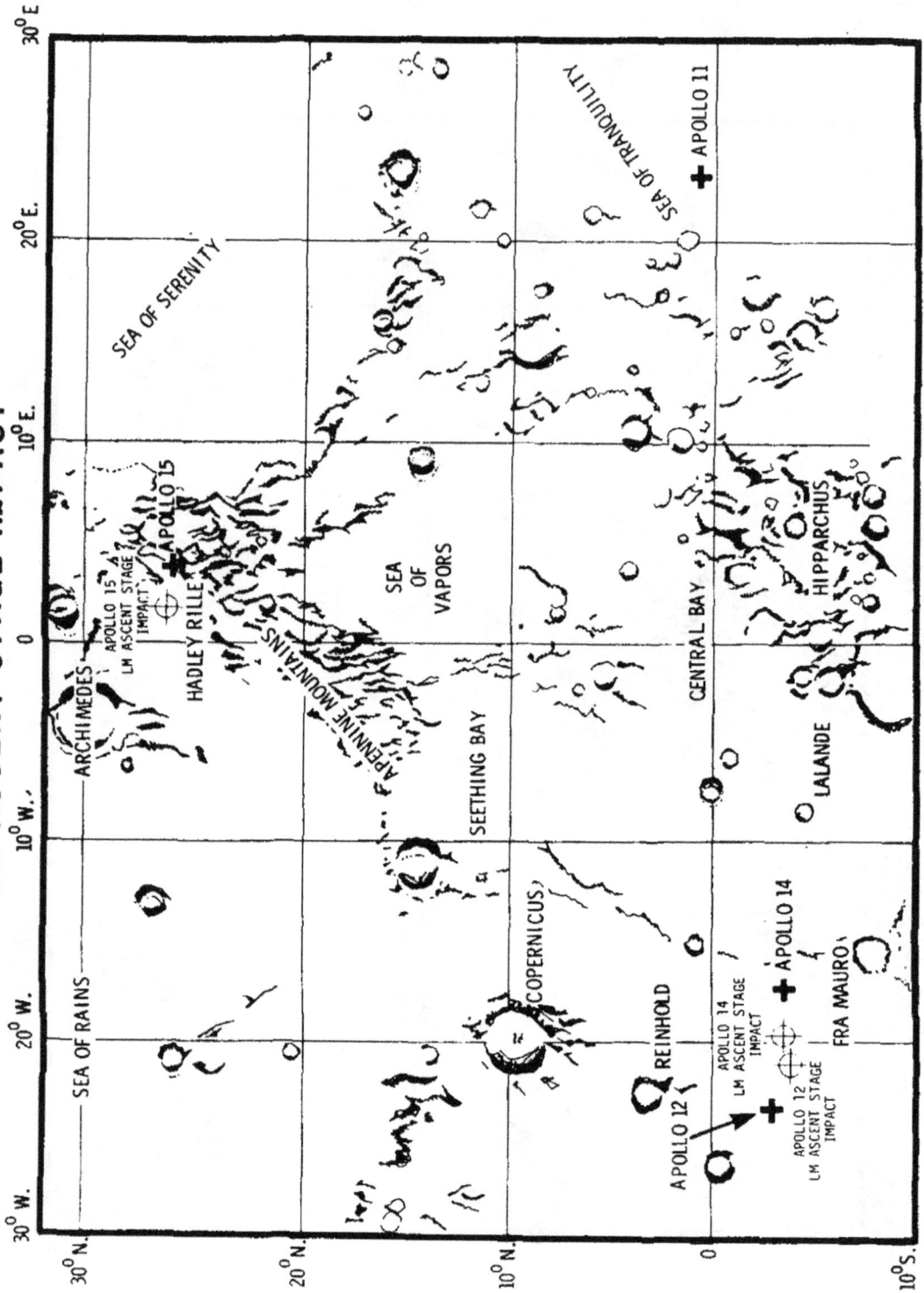

3.   The five-foot diameter hat-shaped thermal shroud covers and helps stabilize the temperature of the sensor assembly. The instrument uses thermostatically controlled heaters to protect it from the extreme cold of the lunar flight.

The Lunar Surface Magnetometer (LSM): The scientific objective of the magnetometer experiment is to measure the magnetic field at the lunar surface. Charged particles and the magnetic field of the solar wind impact directly on the lunar surface. Some of the solar wind particles are absorbed by the surface layer of the Moon. Others may be deflected around the Moon. The electrical properties of the material making up the Moon determine what happens to the magnetic field when it hits the Moon. If the Moon is a perfect insulator the magnetic field will pass through the Moon undisturbed. If there is material present which acts as a conductor, electric currents will flow in the Moon. A small magnetic field of approximately 35 gammas, one thousandth the size of the Earth's field was recorded at the Apollo 12 site. Similar small fields were recorded by the portable magnetometer on Apollo 14.

Two possible models are shown in the next drawing. The electric current carried by the solar wind goes through the Moon and "closes" in the space surrounding the Moon (figure a). This current (E) generates a magnetic field (M) as shown. The magnetic field carried in the solar wind will set up a system of electric currents in the Moon or along the surface. These currents will generate another magnetic field which tries to counteract the solar wind field (figure b). This results in a change in the total magnetic field measured at the lunar surface.

The magnitude of this difference can be determined by independently measuring the magnetic field in the undisturbed solar wind nearby, yet away from the Moon's surface. The value of the magnetic field change at the Moon's surface can be used to deduce information on the electrical properties of the Moon. This, in turn, can be used to better understand the internal temperature of the Moon and contribute to better understanding of the origin and history of the Moon.

The design of the tri-axis flux-gate magnetometer and analysis of experiment data are the responsibility of Dr. Palmer Dyal - NASA/Ames Research Center.

LUNAR MAGNETIC ENVIRONMENT

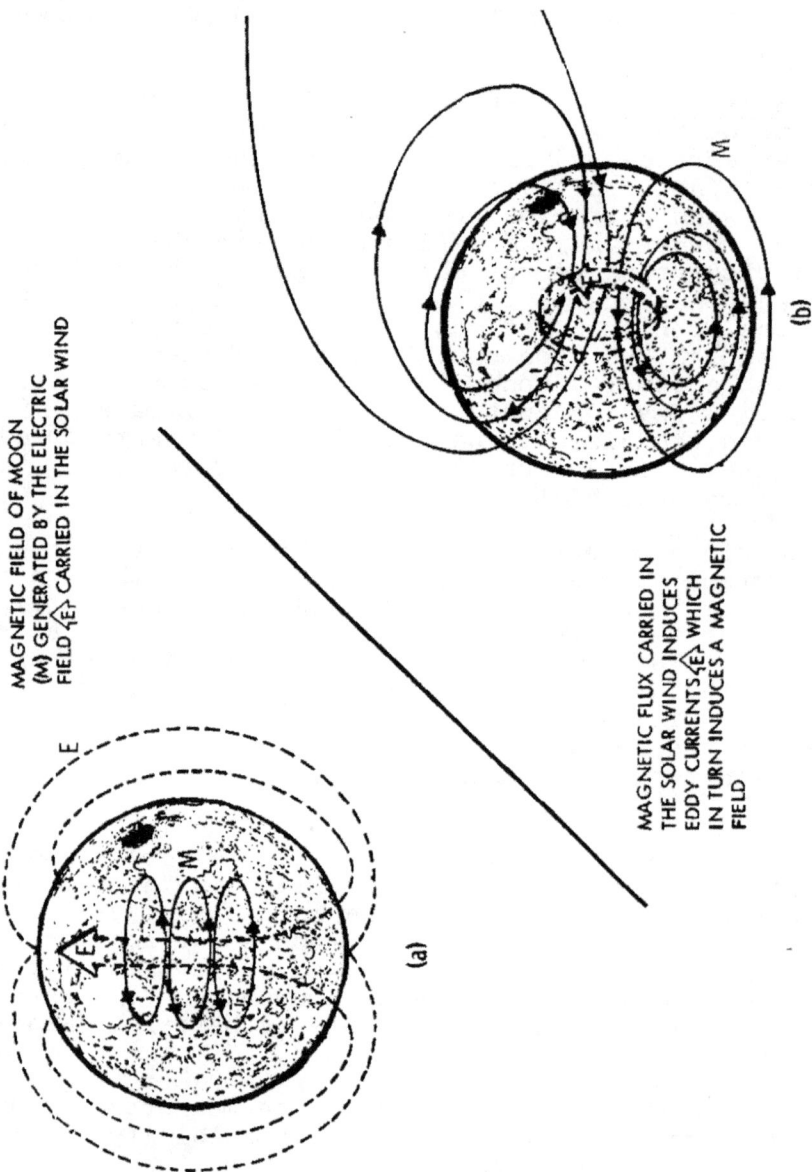

MAGNETIC FIELD OF MOON (M) GENERATED BY THE ELECTRIC FIELD ⟨E⟩ CARRIED IN THE SOLAR WIND

(a)

MAGNETIC FLUX CARRIED IN THE SOLAR WIND INDUCES EDDY CURRENTS ⟨E⟩ WHICH IN TURN INDUCES A MAGNETIC FIELD

(b)

The magnetometer consists of three magnetic sensors aligned in three orthogonal sensing axes, each located at the end of a fiberglass support arm extending from a central structure. This structure houses both the experiment electronics and the electro-mechanical gimbal/flip unit which allows the sensor to be pointed in any direction for site survey and calibration modes. The astronaut aligns the magnetometer experiment to within ± 3 degrees east-west using a shadowgraph on the central structure, and to within ± 3 degrees of the vertical using a bubble level mounted on the Y sensor boom arm.

Size, weight and power are as follows:

Size (inches) deployed	40 high with 60 between sensor heads
Weight (pounds)	17.5
Peak Power Requirements (watts)	
Site Survey Mode	11.5
Scientific Mode	6.2 12.3 (night)
Calibration Mode	10.8

The Magnetometer experiment operates in three modes:

Site Survey Mode -- An initial site survey is performed in each of the three sensing modes for the purpose of locating and identifying any magnetic influences permanently inherent in the deployment site so that they will not affect the interpretation of the LSM sensing of magnetic flux at the lunar surface.

Scientific Mode -- This is the normal operating mode wherein the strength and direction of the lunar magnetic field are measured continuously. The three magnetic sensors provide signal outputs proportional to the incidence of magnetic field components parallel to their respective axes. Each sensor will record the intensity three times per second which is faster than the magnetic field is expected to change. All sensors have the capability to sense over any one of three dynamic ranges with a resolution of 0.2 gammas.

-100 to +100 gamma

-200 to +200 gamma

-400 to +400 gamma

*Gamma is a unit of intensity of a magnetic field. The Earth's magnetic field at the Equator, for example, is 35,000 gamma. The interplanetary magnetic field from the Sun has been recorded at 5 to 10 gamma.

Calibration Mode - This is performed automatically at 12-hour intervals to determine the absolute accuracy of the magnetometer sensors and to correct any drift from their laboratory calibration.

The Solar Wind Spectrometer: The Solar Wind Spectrometer will measure the strength, velocity and directions of the electrons and protons which emanate from the Sun and reach the lunar surface. The solar wind is the major external force working on the Moon's surface. The spectrometer measurements will help interpret the magnetic field of the Moon, the lunar atmosphere and the analysis of lunar samples.

Knowledge of the solar wind will help us understand the origin of the Sun and the physical processes at work on the Sun, i.e., the creation and acceleration of these particles and how they propagate through interplanetary space. It has been calculated that the solar wind puts one kiloton of energy into the Earth's magnetic field every second. This enormous amount of energy influences such Earth processes as the aurora, ionosphere and weather. Although it requires 20 minutes for a kiloton to strike the Moon its effects should be apparent in many ways.

In addition to the Solar Wind Spectrometer, an independent experiment (the Solar Wind Composition Experiment) will collect the gases of the solar wind for return to Earth for analysis.

The design of the spectrometer and the subsequent data analysis are the responsibility of Dr. Conway Snyder of the Jet Propulsion Laboratory.

Seven identical modified Faraday cups (an instrument that traps ionized particles) are used to detect and collect solar wind electrons and protons. One cup is to the vertical, whereas the remaining six cups surround the vertical where the angle between the normals of any two adjacent cups is approximately 60 degrees. Each cup measures the current produced by the charged particle flux entering into it. Since the cups are identical, and if particle flux is equal in each direction, equal current will be produced in each cup. If the flux is not equal in each direction, analysis of the amount of current in the seven cups will determine the variation of particle flow with direction. Also, by successively changing the voltages on the grid of the cup and measuring the corresponding current, complete energy spectra of both electrons and protons in the solar wind are produced.

Data from each cup are processed in the ALSEP data subsystem. The measurement cycle is organized into 16 sequences of 186 ten-bit words. The instrument weighs 12.5 pounds, has an input voltage of about 28.5 volts and has an average input power of about 3.2 watts. The measurement ranges are as follows:

Electrons

| High gain modulation | 10.5 - 1,376 e.v. (electron volts) |
| Low gain modulation | 6.2 - 817 e.v. |

Protons

| High gain modulation | 75 - 9,600 e.v. |
| Low gain modulation | 45 - 5,700 e.v. |

Field of View	6.0 Steradians
Angular Resolution	15 degrees (approximately)
Minimum Flux Detectable	10^6 particles/cm^2/sec

Suprathermal Ion Detector Experiment (SIDE) and Cold Cathode Gauge Experiment: The SIDE will measure flux, composition, energy and velocity of low-energy positive ions and the high-energy solar wind flux of positive ions. Combined with the SIDE is the Cold Cathode Gauge Experiment (CCGE) for measuring the density of the lunar ambient atmosphere and any variations with time or solar activity such atmosphere may have.

Data gathered by the SIDE will yield information on: (1) interaction between ions reaching the Moon from outer space and captured by lunar gravity and those that escape; (2) whether or not secondary ions are generated by ions impacting the lunar surface; (3) whether volcanic processes exist on the Moon; (4) effects of the ambient electric field; (5) loss rate of contaminants left in the landing area by the LM and the crew; and (6) ambient lunar atmosphere pressure.

Dr. John Freeman of Rice University is the SIDE principal investigator, and Dr. Francis B. Johnson of the University of Texas is the CCGE principal investigator.

The SIDE instrument consists of a velocity filter, a low-energy curved-plate analyzer ion detector and a high-energy curved-plate analyzer ion detector housed in a case measuring 15.2 by 4.5 by 13 inches, a wire mesh ground plane, and electronic circuitry to transfer data to the ALSEP central station. The SIDE case rests on folding tripod legs. Dust covers, released by ground command, protect both instruments. Total SIDE weight is 19.6 pounds.

The SIDE and the CCGE connected by a short cable, will be deployed about 55 feet northeast of the ALSEP central station, with the SIDE aligned east or west toward the subearth point and the CCGE orifice aligned along the north-south line with a clear field away from other ALSEP instruments and the LM.

The Cold Cathode Gauge on Apollo 14 is measuring a pressure of 10^{-11} to 10^{-12} torr (where one torr is equal to one millimeter of mercury and 760 millimeters of mercury equal one Earth atmosphere).

Lunar Heat Flow Experiment (HFE): The scientific objective of the Heat Flow experiment is to measure the steady-state heat flow from the lunar interior. Two predicted sources of heat are: (1) original heat at the time of the Moon's formation and (2) radioactivity. Scientists believe that heat could have been generated by the infalling of material and its subsequent compaction as the Moon was formed. Moreover, varying amounts of the radioactive elements uranium, thorium and potassium were found present in the Apollo 11 and 12 lunar samples which if present at depth, would supply significant amounts of heat. No simple way has been devised for relating the contribution of each of these sources to the present rate of heat loss. In addition to temperature, the experiment is capable of measuring the thermal conductivity of the lunar rock material.

The combined measurement of temperature and thermal conductivity gives the net heat flux from the lunar interior through the lunar surface. Similar measurements on Earth have contributed basic information to our understanding of volcanoes, earthquakes and mountain building processes. In conjunction with the seismic and magnetic data obtained on other lunar experiments the values derived from the heat flow measurements will help scientists to build more exact models of the Moon and thereby give us a better understanding of its origin and history.

The Heat Flow experiment consists of instrument probes, electronics and emplacement tool and the lunar surface drill. Each of two probes is connected by a cable to an electronics box which rests on the lunar surface. The electronics, which provide control, monitoring and data processing for the experiment, are connected to the ALSEP central station.

Each probe consists of two identical 20-inch (50 cm) long sections each of which contains a "gradient" sensor bridge, a "ring" sensor bridge and two heaters. Each bridge consists of four platinum resistors mounted in a thin-walled fiberglass cylindrical shell. Adjacent areas of the bridge are located in sensors at opposite ends of the 20-inch fiberglass probe sheath. Gradient bridges consequently measure the temperature difference between two sensor locations.

PROBE PACKAGE CABLE TRAY

ELECTRONICS PACKAGE

PROBE CARRYING PACKAGE (CONTAINS 2 PROBES & EMPLACEMENT TOOL)

SUNSHIELD

THERMAL MASK

REFLECTOR

CABLE BRACKET REMOVED DURING DEPLOYMENT

LUNAR SURFACE

TO ELECTRONICS

RADIATION SHIELD

RING SENSOR (4/PROBE)

GRADIENT SENSOR (INSIDE) 4/PROBE

HEATER COILS (OUTSIDE)

PROBE STOP

THERMOCOUPLES (4) 25.6, 45.3 & 65.0 IN. ABOVE PROBE

RADIATION SHIELD

FLEXIBLE SPRING

PROBE

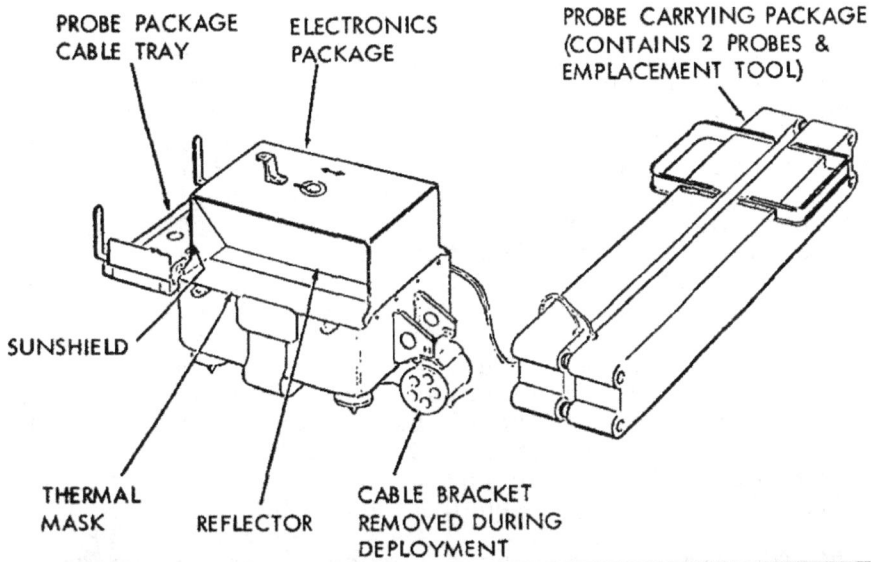

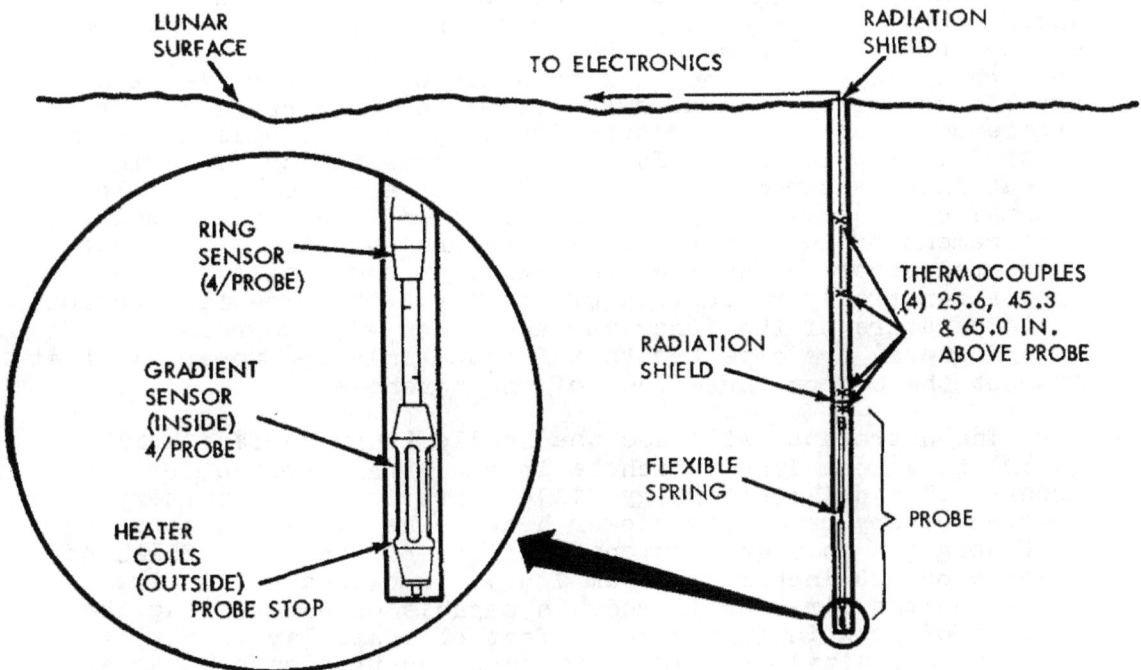

HEAT FLOW EXPERIMENT

In thermal conductivity measurements at very low values a heater surrounding the gradient sensor is energized with 0.002 watts and the gradient sensor values monitored. The rise in temperature of the gradient sensor is a function of thermal conductivity of the surrounding lunar material. For higher range of values, the heater is energized at 0.5 watts of heat and monitored by a ring sensor. The rate of temperature rise, monitored by the ring sensor is a function of the thermal conductivity of the surrounding lunar material. The ring sensor, approximately four inches from the heater, is also a platinum resistor. A total of eight thermal conductivity measurements can be made. The thermal conductivity mode of the experiment will be implemented about 20 days (500 hours) after deployment. This is to allow sufficient time for the perturbing effects of drilling and emplacing the probe in the borehole to decay; i.e., for the probe and casings to come to equilibrium with the lunar subsurface.

A 30-foot (10-meter) cable connects each probe to the electronics box. In the upper six feet of the borehole the cable contains four evenly spaced thermocouples: at the top of the probe; at 26 inches (65 cm), 45 inches (115 cm), and 66 inches (165 cm). The thermocouples will measure temperature transients propagating downward from the lunar surface. The reference junction temperature for each thermocouple is located in the electronics box. In fact, the feasibility of making a heat flow measurement depends to a large degree on the low thermal conductivity of the lunar surface layer, the regolith. Measurement of lunar surface temperature variations by Earth-based telescopes as well as the Surveyor and Apollo missions show a remarkably rapid rate of cooling. The wide fluctuations in temperature of the lunar surface (from -250 degrees F to +250 degrees) are expected to influence only the upper six feet and not the bottom three feet of the borehole.

The astronauts will use the Apollo Lunar Surface Drill (ALSD) to make a lined borehole in the lunar surface for the probes. The drilling energy will be provided by a battery-powered rotary percussive power head. The drill rod consists of fiberglass tubular sections reinforced with boron filaments (each about 20 inches or 50 cm long). A closed drill bit, placed on the first drill rod, is capable of penetrating the variety of rock including three feet of vesicular basalt (40 per cent porosity). As lunar surface penetration progresses, additional drill rod sections will be connected to the drill string. The drill string is left in place to serve as a hole casing.

An emplacement tool is used by the astronaut to insert the probe to full depth. Alignment springs position the probe within the casing and assure a well-defined radiative coupling between the probe and the borehole. Radiation shields on the hole prevent direct sunlight from reaching the bottom of the hole.

The astronaut will drill a third hole near the HFE and obtain cores of lunar material for subsequent analysis of thermal properties. Total available core length is 100 inches.

Heat flow experiments, design and data analysis are the responsibility of Dr. Marcus Langseth of the Lamont-Doherty Geological Observatory.

Lunar Dust Detector Experiment: Separates and measures high-energy radiation damage to three solar cells, measures reduction of solar cell output due to dust accumulation and measures reflected infrared energy and temperatures for computation of lunar surface temperatures. A sensor package is mounted on the ALSEP central station sunshield and a printed circuit board inside the central station monitors the data subsystem power distribution unit. Principal investigator: James R. Bates, NASA Manned Spacecraft Center.

ALSEP Central Station: The ALSEP Central Station serves as a power-distribution and data-handling point for experiments carried on each version of the ALSEP. Central Station components are the data subsystem, helical antenna, experiment electronics, power conditioning unit and dust detector. The Central Station is deployed after other experiment instruments are unstowed from the pallet.

The Central Station data subsystem receives and decodes uplink commands, times and controls experiments, collects and transmits scientific and engineering data downlink, and controls the electrical power subsystem through the power distribution and signal conditioner.

The modified axial helix S-band antenna receives and transmits a right-hand circularly-polarized signal. The antenna is manually aimed with a two-gimbal azimuth/elevation aiming mechanism. A dust detector on the Central Station, composed of three solar cells, measures the accumulation of lunar dust on ALSEP instruments.

The ALSEP electrical power subsystem draws electrical power from a SNAP-27 (Systems for Nuclear Auxiliary Power) radioisotope thermoelectric generator.

Laser Ranging Retro-Reflector (LRRR) Experiment: The LRRR will permit long-term measurements of the Earth-Moon distance by acting as a passive target for laser beams directed from observatories on Earth. Data gathered from these measurements of the round trip time for a laser beam will be used in the study of fluctuations in the Earth's rotation rate, wobbling motions of the Earth on its axis, the Moon's size and orbital shape, and the possibility of a slow decrease in the gravitational constant "G". Dr. James Faller of Wesleyan University, Middletown, CT, is LRRR principal investigator.

The LRRR is a square array of 300 fused silica reflector cubes mounted in an adjustable support structure which will be aimed toward Earth by the crew during deployment. Each cube reflects light beams back in absolute parallelism in the same direction from which they came.

By timing the round trip time for a laser pulse to reach the LRRR and return, observatories on Earth can calculate the exact distance from the observatory to the LRRR location within a tolerance of ±6 cm (or one foot). A 100-cube LRRR was deployed at Tranquillity Base by the Apollo 11, and at Fra Mauro by the Apollo 14 crew. The goal is to set up LRRRs at three lunar locations to establish absolute control points in the study of Moon motion.

Solar Wind Composition Experiment:(SWC): The scientific objective of the solar wind composition experiment is to determine the elemental and isotopic composition of the noble gases in the solar wind.

As in Apollos 11, 12, and 14, the SWC detector will be deployed on the lunar surface and brought back to Earth by the crew. The detector will be exposed to the solar wind flux for 45 hours compared to 21 hours on Apollo 14, 17 hours on Apollo 12, and two hours on Apollo 11.

The solar wind detector consists of an aluminum foil four square feet in area and about 0.5 mils thick rimmed by Teflon for resistance to tearing during deployment. A staff and yard arrangement will be used to deploy the foil and to maintain the foil approximately perpendicular to the solar wind flux. Solar wind particles will penetrate into the foil, allowing cosmic rays to pass through. The particles will be firmly trapped at a depth of several hundred atomic layers. After exposure on the lunar surface, the foil is rolled up and returned to Earth. Professor Johannes Geiss, University of Berne, Switzerland, is principal investigator.

SNAP-27 -- Power Source for ALSEP: A SNAP-27 unit,
similar to two others on the Moon, will provide power for the
ALSEP package. SNAP-27 is one of a series of radioisotope
thermoelectric generators, or atomic batteries developed by
the Atomic Energy Commission under its space SNAP program.
The SNAP (Systems for Nuclear Auxiliary Power) program is
directed at development of generators and reactors for use
in space, on land, and in the sea.

While nuclear heaters were used in the seismometer
package on Apollo 11, SNAP-27 on Apollo 12 marked the first
use of a nuclear electrical power system on the Moon. The use
of SNAP-27 on Apollo 14 marked the second use of such a unit
on the Moon. The first unit has already surpassed its one-year
design life by eight months, thereby allowing the simultaneous
operation of two instrument stations on the Moon.

The basic SNAP-27 unit is designed to produce at least
63.5 electrical watts of power. The SNAP-27 unit is a cylindrical generator, fueled with the radioisotope plutonium-238.
It is about 18 inches high and 16 inches in diameter, including
the heat radiating fins. The generator, making maximum use of
the lightweight material beryllium, weighs about 28 pounds
unfueled.

The fuel capsule, made of a superalloy material, is
16.5 inches long and 2.5 inches in diameter. It weighs about
15.5 pounds, of which 8.36 pounds represent fuel. The plu-
tonium-238 fuel is fully oxidized and is chemically and bio-
logically inert.

The rugged fuel capsule is contained within a graphite
fuel cask from launch through lunar landing. The cask is designed to provide reentry heating protection and added contain-
ment for the fuel capsule in the event of an aborted mission.
The cylindrical cask with hemispherical ends includes a primary
graphite heat shield, a secondary beryllium thermal shield,
and a fuel capsule support structure. The cask is 23 inches
long and eight inches in diameter and weighs about 24.5 pounds.
With the fuel capsule installed, it weighs about 40 pounds. It
is mounted on the lunar module descent stage.

Once the lunar module is on the Moon, an Apollo
astronaut will remove the fuel capsule from the cask and insert it into the SNAP-27 generator which will have been placed
on the lunar surface near the module.

The spontaneous radioactive decay of the plutonium-238
within the fuel capsule generates heat which is converted directly
into electrical energy -- at least 63.5 watts. There are no
moving parts.

The unique properties of plutonium-238 make it an excellent isotope for use in space nuclear generators. At the end of almost 90 years, plutonium-238 is still supplying half of its original heat. In the decay process, plutonium-238 emits mainly the nuclei of helium (alpha radiation), a very mild type of radiation with a short emission range.

Before the use of the SNAP-27 system in the Apollo program was authorized, a thorough review was conducted to assure the health and safety of personnel involved in the launch and of the general public. Extensive safety analyses and tests were conducted which demonstrated that the fuel would be safely contained under almost all credible accident conditions.

<u>Lunar Geology Investigation</u>: The Hadley/Apennines site was
selected for multiple objectives: 1) the Apennine Mountain
front, 2) the sinuous Hadley Rille, 3) the dark mare material
of Palus Putredinis, 4) the complex of domical hills in the
mare, and 5) the arrowhead-shaped crater cluster.

The Apennine Mountain front forms the arcuate southeastern rim of Mare Imbrium. It borders Palus Putredinis
and, in the area of the site, it rises 12,000 feet above
the surrounding mare. The Apennine Mountain front is
believed to have been exposed at the time of the excavation
of the giant Imbrium basin. The cratering event must have,
therefore, exposed materials which are pre-Imbrian in age.
Examination and collection of this ancient material as well
as deep-seated Imbrium ejecta are the prime objectives of
the mission. This will be accomplished during the first and
second EVA's when Scott and Irwin will select samples from
the foot of the mountain scarp and from the ejecta blankets of
craters which excavate mountain materials.

The second important objective of the mission is to study
and sample the Hadley Rille, which runs parallel to the
Apennine Mountain front and incises the Palus Putredinis mare
material. The rille is a sinuous or meandering channel, much
like a river gorge on Earth. It displays a V-shaped cross
section, with an average slope of about 25°. The rille
originates in an elongate depression near the base of the
mountain some 40 miles south of the site. In the vicinity
of the landing site, the rille is about one mile wide and
1,200 feet deep. The origin of the rille is not known and
it is hoped that samples collected at its rim and high resolution photographs of its walls will unravel its mode of formation

The third objective of the mission is to study and sample
the reasonably flat mare material of Palus Putredinis on which
the LM will land. This mare material is dark and preliminary
studies of crater distribution indicate that this mare surface
is younger than that visited on Apollo 11, and probably is
closer to the age of the Apollo 12 mare site. Systematic
sampling of this surface unit will be done by visiting craters
which have penetrated it to various depths. The Apollo 15
crew will use the standard lunar hand tools used on past
missions for sampling. However, the hand tool carrier will be
mounted on the lunar roving vehicle (LRV).

A complex of domical structures about 5 km north of the landing site constitutes another objective of the mission. The hills may be made of volcanic domes superposed on the surrounding mare or buried domical structures thinly covered by the mare-like material. Amont the hills are large craters which have excavated subsurface material for sampling as well as interesting linear depressions and ridges.

The fifth sampling objective of the mission is a cluster of craters which forms an arrowhead-shaped pattern. This crater cluster is aligned along a ray from the crater Autolycus, over 100 miles northwest of the site. It is believed to be made of secondary craters from Autolycus ejecta and offers a good opportunity to study the features and perhaps sample material which originated at Autolycus.

Planned sampling sites for the mission allow therefore a thorough investigation of a variety of features. The most important of all features in the area is the Apennine Mountain front, where samples of the oldest exposed rocks on the Moon may be obtained.

In addition to planned sampling sites, Scott and Irwin will select other sites for gathering, observing and photo-graphing geological samples. Both men will use chest-mounted Hasselblad electric data cameras for documenting the samples in their natural state. Core tube samples will also be retrieved for geological and geochemical investigation.

Soil Mechanics: Mechanical properties of the lunar soil, surface and subsurface, will be investigated through trenching at various locations, and through use of the self-recording penetrometer equipped with interchangeable cones of various sizes and a load plate. This experiment will be documented with the electric Hasselblad and the 16mm data acquisition camera.

Lunar Orbital Science

Service Module Sector 1, heretofore vacant except for a third cyrogenic oxygen tank added after the Apollo 13 incident, houses the Scientific Instrument Module (SIM) bay on the Apollo J missions.

Eight experiments are carried in the SIM bay: X-ray fluorescence detector, gamma ray spectrometer, alpha-particle spectrometer, panoramic camera, 3-inch mapping camera, laser altimeter and dual-beam mass spectrometer; a subsatellite carrying three integral experiments (particle detectors, magnetometer and S-Band transponder) comprise the eighth SIM bay experiment and will be jettisoned into lunar orbit.

Gamma-Ray Spectrometer: On a 25-foot extendable boom. Measures chemical composition of lunar surface in conjunction with X-ray and alpha-particle experiments to gain a compositional "map" of the lunar surface ground track. Detects natural and cosmic rays, induced gamma radioactivity and will operate on Moon's dark and light sides. Additionally, the experiment will be extended in transearth coast to measure the radiation flux in cislunar space and record a spectrum of cosmological gamma-ray flux. The device can measure energy ranges between 0.1 to 10 million electron volts. The extendable boom is controllable from the command module cabin. Principal investigator: Dr. James R. Arnold, University of California at San Diego.

X-Ray Fluorescence Spectrometer: Second of the geochemical experiment trio for measuring the composition of the lunar surface from orbit, and detects X-ray fluorescence caused by solar X-ray interaction with the Moon. It will analyze the sunlit portion of the Moon. The experiment will measure the galactic X-ray flux during transearth coast. The device shares a compartment on the SIM bay lower shelf with the alpha-particle experiment, and the protective door may be opened and closed from the command module cabin. Principal Investigator: Dr. Isidore Adler, NASA Goddard Space Flight Center, Greenbelt, MD.

Alpha-Particle Spectrometer: Measures mono-energetic alpha-particles emitted from the lunar crust and fissures as products of radon gas isotopes in the energy range of 4.7 to 9.3 million electron volts. The sensor is made up of an array of 10 silicon surface barrier detectors. The experiment will construct a "map" of lunar surface alpha-particle emissions along the orbital track and is not constrained by solar illumination. It will also measure deep-space alpha-particle background emissions in lunar orbit and in transearth coast. Protective door operation is controlled from the cabin. Principal investigator: Dr. Paul Gorenstein, American Science and Engineering, Inc., Cambridge, MA.

Mass Spectrometer: Measures composition and distribution of
the ambient lunar atmosphere, identifies active lunar sources
of volatiles, pinpoints contamination in the lunar atmosphere.
The sunset and sunrise terminators are of special interest,
since they are predicted to be regions of concentration of
certain gases. Measurements over at least five lunar revolu-
tions are desired. The mass spectrometer is on a 24-foot
extendable boom. The instrument can identify species from 12
to 28 atomic mass units (AMU) with No. 1 ion counter, and
28-66 AMU with No. 2 counter. Principal investigator:
Dr. John H. Hoffman, University of Texas at Dallas.

24-inch Panoramic Camera (SM orbital photo task): Gathers
stereo and high-resolution (1 meter) photographs of the lunar
surface from orbit. The camera produces an image size of
15 x 180 nm with a field of view 11° downtrack and 108° cross
track. The rotating lens system can be stowed face-inward to
avoid contamination during effluent dumps and thruster firings.
The 72-pound film cassette of 1,650 frames will be retrieved
by the command module pilot during a transearth coast EVA.
The 24-inch camera works in conjunction with the 3-inch
mapping camera and the laser altimeter to gain data to construct
a comprehensive map of the lunar surface ground track flown
by this mission---about 1.16 million square miles, or 8 percent
of the lunar surface.

3-inch Mapping Camera: Combines 20-meter resolution terrain
mapping photography on five-inch film with 3-inch focal length
lens with stellar camera shooting the star field on 35mm film
simultaneously at 96° from the surface camera optical axis.
The stellar photos allow accurate correlation of mapping photography postflight by comparing simultaneous star field photos
with lunar surface photos of the nadir (straight down). Additionally, the stellar camera provides pointing vectors for the
laser altimeter during darkside passes. The 3-inch f4.5 mapping
camera metric lens covers a 74° square field of view, or 92x92 nm
from 60 nm altitude. The stellar camera is fitted with a 3-inch
f/2.8 lens covering a 24° field with cone flats. The 23-pound
film cassette containing mapping camera film (3,600) frames) and
the stellar camera film will be retrieved during the same EVA
described in the panorama camera discussion. The Apollo Orbital
Science Photographic Team is headed by Frederick J. Doyle of the
U.S. Geological Survey, McLean, VA.

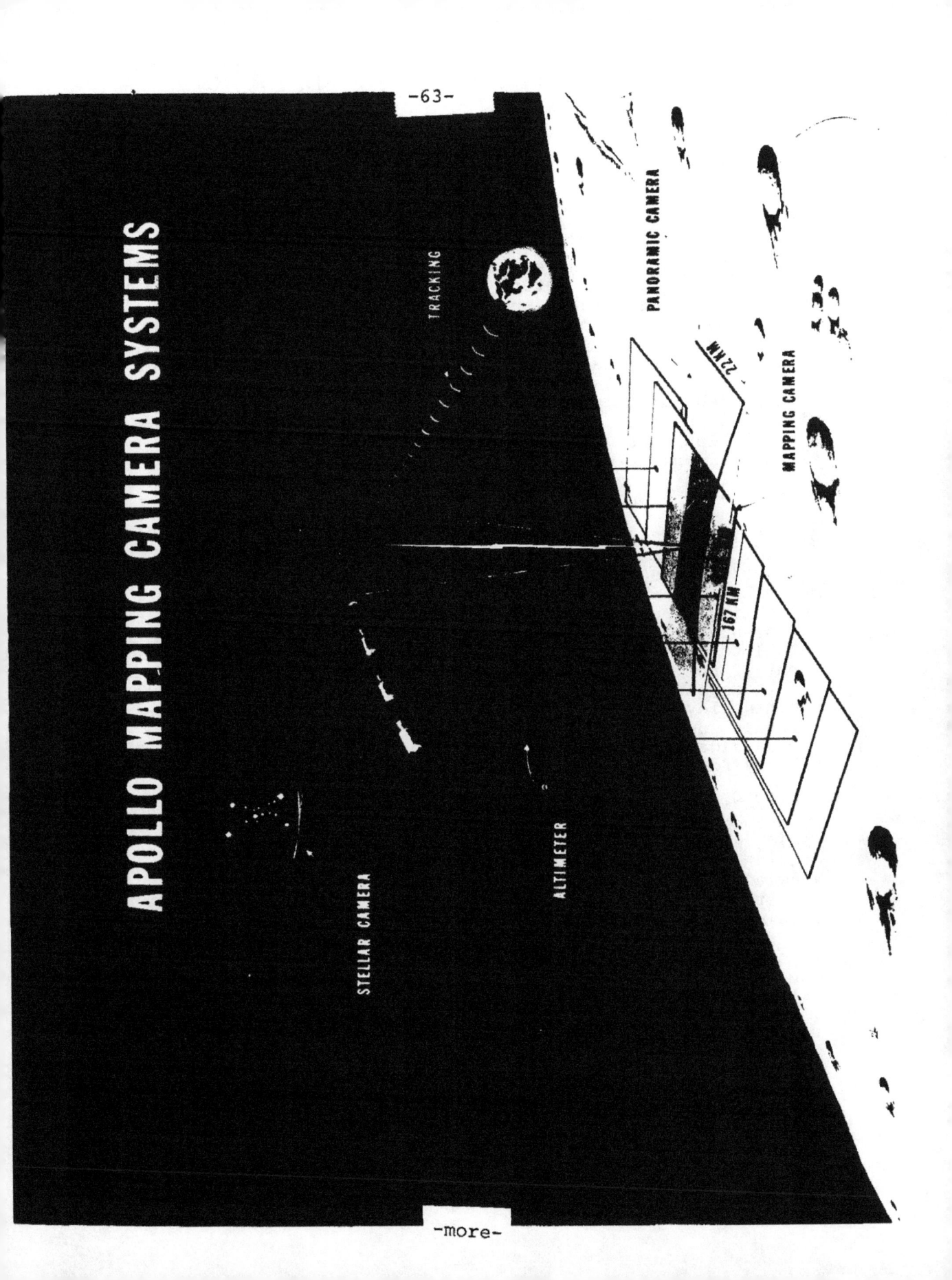

APOLLO MAPPING CAMERA SYSTEMS

STELLAR CAMERA

ALTIMETER

TRACKING

222 NM

167 KM

PANORAMIC CAMERA

MAPPING CAMERA

Laser Altimeter: Measures spacecraft altitude above the lunar surface to within one meter. Instrument is boresighted with 3-inch mapping camera to provide altitude correlation data for the mapping camera as well as the 24-inch panoramic camera. When the mapping camera is running, the laser altimeter automatically fires a laser pulse corresponding to mid-frame ranging to the surface for each frame. The laser light source is a pulsed ruby laser operating at 6,943 angstroms, and 200-millijoule pulses of 10 nanoseconds duration. The laser has a repetition rate up to 3.75 pulses per minute. The laser altimeter working group of the Apollo Orbital Science Photographic Team is headed by Dr. William M. Kaula of the UCLA Institute of Geophysics and Planetary Physics.

Subsatellite: Ejected into lunar orbit from the SIM bay and carries three experiments: S-Band Transponder, Particle Shadows/Boundary Layer Experiment, and Subsatellite Magnetometer Experiment. The subsatellite is housed in a container resembling a rural mailbox, and when deployed is spring-ejected out-of-plane at 4 fps with a spin rate of 140 rpm. After the satellite booms are deployed, the spin rate is stabilized at about 12 rpm. The subsatellite is 31 inches long, has a 14-inch hexagonal diameter and weighs 78.5 pounds. The folded booms deploy to a length of five feet. Subsatellite electrical power is supplied by a solar cell array outputting 25 watts for dayside operation and a rechargeable silver-cadmium battery for nightside passes.

Experiments carried aboard the subsatellite are: S-Band transponder for gathering data on the lunar gravitational field, especially gravitational anomalies such as the so-called mascons; Particle Shadows/Boundary Layer for gaining knowledge of the formation and dynamics of the Earth's magnetosphere, interaction of plasmas with the Moon and the physics of solar flares using telescope particle detectors and spherical electrostatic particle detectors; and Subsatellite Magnetometer for gathering physical and electrical property data on the Moon and of plasma interaction with the Moon using a biaxial flux-gate magnetometer deployed on one of the three five-foot folding booms. Principal investigators for the subsatellite experiments are: Particle Shadows/Boundary Layer, Dr. Kinsey A. Anderson, University of California Berkeley; Magnetometer, Dr. Paul J. Coleman, UCLA; and S-Band Transponder, Mr. William Sjogren, Jet Propulsion Laboratory.

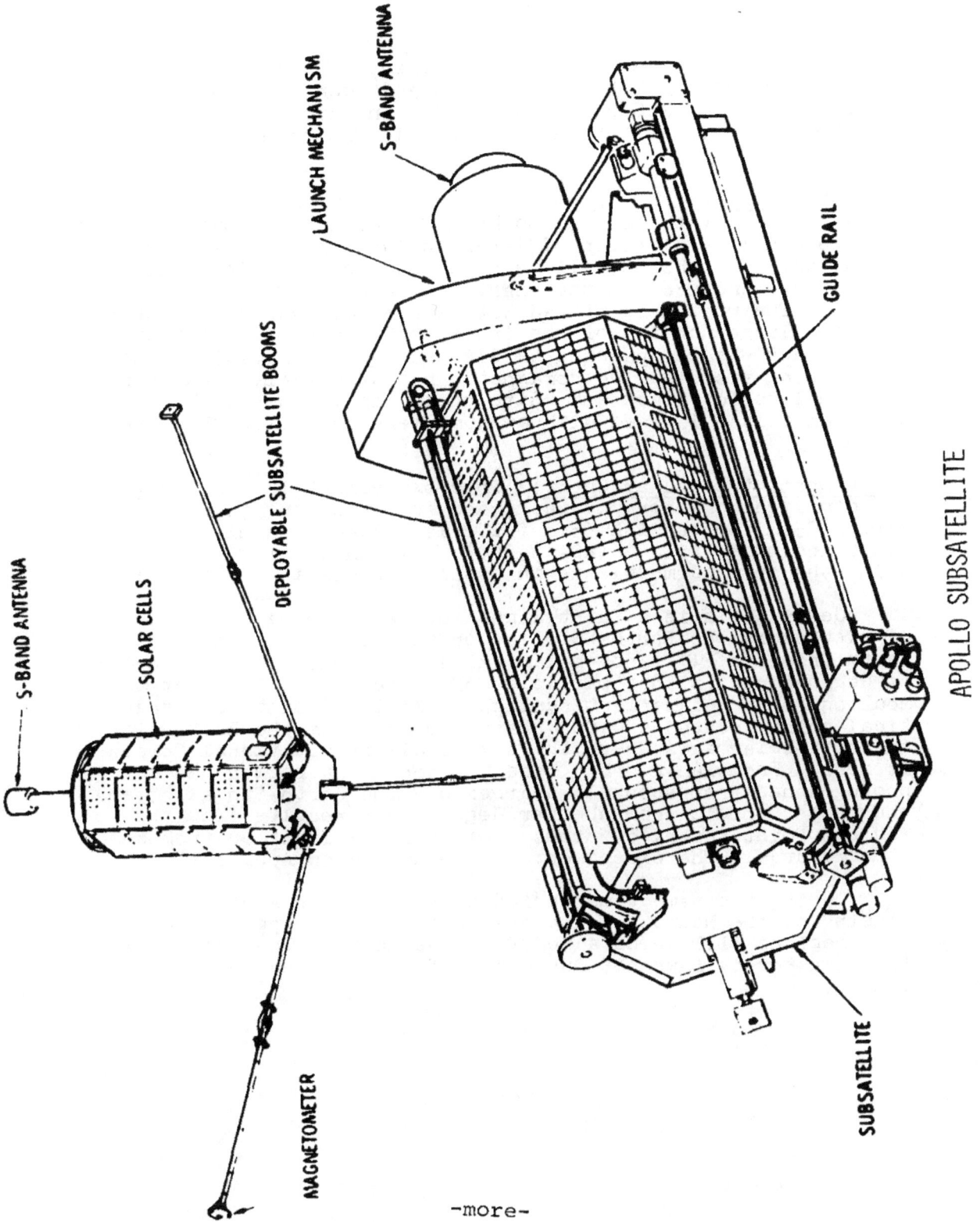

LAUNCH MECHANISM

S-BAND ANTENNA

GUIDE RAIL

DEPLOYABLE SUBSATELLITE BOOMS

S-BAND ANTENNA

SOLAR CELLS

MAGNETOMETER

SUBSATELLITE

APOLLO SUBSATELLITE

Other CSM orbital science experiments and tasks not in the SIM bay include UV Photography-Earth and Moon, Gegenschein from Lunar Oribt, CSM/LM S-Band Transponder in addition to the Subsatellite, Bistatic Radar, and Apollo Window Meteoroid experiments.

UV Photography-Earth and Moon: Aimed toward gathering ultra-violet photos of the Earth and Moon for planetary atmosphere studies and investigation of lunar surface short wavelength radiation. The photos will be made with an electric Hassel-blad bracket mounted in the right side window of the command module. The window is fitted with a special quartz pane that passes a large portion of the incident UV spectrum. A four-filter pack---three passing UV electromagnetic radiation and one passing visible electromagnetic radiation---is used with a 105mm lens for black and white photography; the visible spectrum filter is used with an 80mm lens for color UV photography.

Gegenschein from Lunar Orbit: This experiment is similar to the dim light photography task, and involves long exposures with a 35mm camera with 55mm f/1.2 lens camera on high speed black and white film (ASA 6,000). All photos must be made while the command module is in total darkness in lunar orbit.

Gegenschein is a faint light source covering a 20° field of view along the Earth-Sun line on the opposite side of the Earth from the Sun (anti-solar axis). One theory on the origin of Gegenschein is that particles of matter are trapped at the Moulton Point and reflect sunlight. Moulton Point is a theoretical point located 940,000 statute miles from the Earth along the anti-solar axis where the sum of all gravitational forces is zero. From lunar orbit, the Moulton Point region can be photographed from about 15 degrees off the Earth-Sun axis, and the photos should show whether Gegenschein results from the Moulton Point theory or stems from zodiacal light or from some other source. The experiment was conducted on Apollo 14.

During the same time period that photographs of the Gegenschein and the Moulton Point are taken, photographs of the same regions will be obtained from the Earth. The principal investigator is Lawrence Dunkelman of the Goddard Space Flight Center.

CSM/LM S-Band Transponder: The objective of this experiment is to detect variations in lunar gravity along the lunar surface track. These anomalies in gravity result in minute perturbations of the spacecraft motion and are indicative of magnitude and location of mass concentrations on the Moon. The Manned Space Flight Network (MSFN) and the Deep Space Network (DSN) will obtain and record S-band Doppler tracking measurements from the docked CSM/LM and the undocked CSM while in lunar orbit; S-band Doppler tracking measurements of the LM during non-powered portions of the lunar descent; and S-band Doppler tracking measurements of the LM ascent stage during non-powered portions of the descent for lunar impact. The CSM and LM S-band Transponders will be operated during the experiment period. The experiment was conducted on Apollo 14.

S-band Doppler tracking data have been analyzed from the Lunar Orbiter missions and definite gravity variations were detected. These results showed the existence of mass concentrations (mascons) in the ringed maria. Confirmation of these results has been obtained with Apollo tracking data.

With appropriate spacecraft orbital geometry much more scientific information can be gathered on the lunar gravitational field. The CSM and/or LM in low-altitude orbits can provide new detailed information on local gravity anomalies. These data can also be used in conjunction with high-altitude data to possibly provide some description on the size and shape of the perturbing masses. Correlation of these data with photographic and other scientific records will give a more complete picture of the lunar environment and support future lunar activities. Inclusion of these results is pertinent to any theory of the origin of the Moon and the study of the lunar subsurface structure. There is also the additional benefit of obtaining better navigational capabilities for future lunar missions in that an improved gravity model will be known. William Sjogren, Jet Propulsion Laboratory, Pasadena, California, is principal investigator.

Bistatic Radar Experiment: The downlink Bistatic Radar Experiment seeks to measure the electromagnetic properties of the lunar surface by monitoring that portion of the spacecraft telemetry and communications beacons which are reflected from the Moon.

The CSM S-band telemetry beacon (f = 2.2875 Gigahertz), the VHF voice communications link (f = 259.7 megahertz), and the spacecraft omni-directional and high gain antennas are used in the experiment. The spacecraft is oriented so that the radio beacon is incident on the lunar surface and is successively reoriented so that the angle at which the signal intersects the lunar surface is varied. The radio signal is reflected from the surface and is monitored on Earth. The strength of the reflected signal will vary as the angle at which it intersects the surface is varied.

By measuring the reflected signal strength as a function of angle of incidence on the lunar surface, the electromagnetic properties of the surface can be determined. The angle at which the reflected signal strength is a minimum is known as the Brewster Angle and determines the dielectric constant. The reflected signals can also be analyzed for data on lunar surface roughness and surface electrical conductivity.

The S-band signal will primarily provide data on the surface. However, the VHF signal is expected to penetrate the gardened debris layer (regolith) of the Moon and be reflected from the underlying rock strata. The reflected VHF signal will then provide information on the depth of the regolith over the Moon.

The S-band BRE signal will be monitored by the 210-foot antenna at the Goldstone, California, site and the VHF portion of the BRE signal will be monitored by the 150-foot antenna at the Stanford Research Institute in California. The experiment was flown on Apollo 14.

Lunar Bistatic Radar Experiments were also performed using the telemetry beacons from the unmanned Lunar Orbiter I in 1966 and from Explorer 35 in 1967. Taylor Howard, Stanford University, is the principal investigator.

Apollo Window Meteoroid: A passive experiment in which command module windows are scanned under high magnification pre- and postflight for evidence of meteoroid cratering flux of one-trillionth gram or larger. Such particle flux may be a factor in degradation of surfaces exposed to space environment. Principal investigator: Burton Cour-Palais, NASA Manned Spacecraft Center.

Composite Casting Demonstration

The Composite Casting technical demonstration performed on the Apollo 14 mission will be carried again on
Apollo 15 to perform more tests on the effects of weight-
lessness on the solidification of alloys, intermetallic
compounds, and reinforced composite materials. Ten samples
will be processed, of which two will be directionally solid-
ified samples of the indium-bismuth eutectic alloy, four
will comprise attempts to make single crystals of the indium-
bismuth intermetallic compound InBi, and four will be models
of composite materials using various types of solid fibers
and particles in matrices of the indium-bismuth eutectic
alloy.

Hardware for the demonstration will include ten welded
aluminum capsules containing the samples, a low-powered electrical resistance heater used to melt the samples, and a
storage box which also serves as a heat sink for directional
solidification. The entire demonstration package will weigh
about ten pounds. Each of the sample capsules is 3.5 inches
long and 7/8-inch in diameter. The heater unit is
cylindrical, with capped openings on its top and bottom, and
is operated from a 28-volt D.C. supply. The storage box is
4.25 by 5 by 3.5 inches.

To process the samples, the astronauts will insert
the capsules one at a time into the heater, apply power for
a prescribed time to melt the sample material, turn off the
heater, and then either allow the assembly to cool without
further attention or, in some cases, mount the heater on the
storage box heat sink to cool. Individual samples will take
from 45 to 105 minutes to process, depending on the material.
All ten samples may not be processed; the deciding factor will
be how much free time the astronauts have to operate the
apparatus during transearth coast phase of the mission. No
data will be taken on the samples in flight.

The returned samples will be evaluated on the ground
by metallurgical, chemical, and physical tests. These results
will be used in conjunction with those already obtained on
the Apollo 14 samples to assess the prospects for further
metallurgical research and eventual product manufacturing in
space.

The demonstration hardware was built at NASA's Marshall
Space Flight Center in Huntsville, Ala.

Engineering/Operational Objectives

In addition to the lunar surface and lunar orbital experiments, there are several test objectives in the Apollo 15 mission aimed toward evaluation of new hardware from an operational or performance standpoint. These test objectives are:

*Lunar Rover Vehicle Evaluation--an assessment of the LRV's performance and handling characteristics in the lunar environment.

*EVA Communications with the lunar communications relay unit/ground commanded television assembly (LCRU/GCTA)--has the objective of demonstrating that the LCRU is capable of relaying two-way communications when the crew is beyond line-of-site from the LM, and that the GCTA can be controlled from the ground for television coverage of EVAs.

*EMU Assessment on Lunar Surface--an evaluation of the improved Apollo spacesuit (A7LB) and the -7 portable life support system (PLSS), both of which are being used for the first time on Apollo 15. The suit modifications allow greater crew mobility, and the later model PLSS allows a longer EVA stay time because of increased consumables.

*Landing Gear Performance of Modified LM--measurements of the LM landing gear stroking under a heavier load caused by J-mission modifications and additons to the basic LM structure--about 1,570 pounds over H-mission LMs.

*SIM Thermal Data--measurement of the thermal responses of the SIM Bay and the experiments stowed in the bay, and the effect of the bay upon the rest of the service module.

*SIM Bay Inspection During EVA--evaluation of the effects of SIM bay door jettison, detect any SIM bay contamination, and evaluate equipment and techniques for EVA retrieval of film cassettes.

*SIM Door Jettison Evaluation--an engineering evaluation of the SIM door jettison mechanisms and the effects of jettison on the CSM.

*LM Descent Engine Performance--evaluation of the descent engine with lengthened engine skirt, longer burn time, and new thrust chamber material.

APOLLO LUNAR HAND TOOLS

Special Environmental Container - The special environmental sample is collected in a carefully selected area and sealed in a special container which will retain a high vacuum. The container is opened in the lunar receiving laboratory (LRL) where it will provide scientists the opportunity to study lunar material in its original environment.

Extension handle - This tool is of aluminum alloy tubing with a malleable stainless steel cap designed to be used as an anvil surface. The handle is designed to be used as an extension for several other tools and to permit their use without requiring the astronaut to kneel or bend down. The handle is approximately 30 inches long and one inch in diameter. The handle contains the female half of a quick disconnect fitting designed to resist compression, tension, torsion or a combination of these loads.

Nine core tubes - These tubes are designed to be driven or augered into loose gravel, sandy material or into soft rock such as feather rock or pumice. They are about 15 inches in length and one inch in diameter and are made of aluminum tubing. Each tube is supplied with a removable non-serrated cutting edge and a screw-on cap incorporating a metal-to-metal crush seal which replaces the cutting edge. The upper end of each tube is sealed and designed to be used with the extension handle or as an anvil. Incorporated into each tube is a spring device to retain loose materials in the tube.

Adjustable Sampling Scoop - Similar to a garden scoop, the device is used for gathering sand or dust too small for the rake or tongs. The stainless steel pan is adjustable from 55 to 90 degrees. The handle is compatible with the extension handle.

Sampling hammer - This tool serves three functions, as a sampling hammer, as a pick or mattock and as a hammer to drive the core tubes or scoop. The head has a small hammer face on one end, a broad horizontal blade on the other, and large hammering flats on the sides. The handle is 14 inches long and is made of formed tubular aluminum. The hammer has on its lower end a quick-disconnect to allow attachment to the extension handle for use as a hoe. The head weight has been increased to provide more impact force.

Collection Bags - Two types of bags are provided for collecting lunar surface samples: the sample collection bag with pockets for holding core tubes, the special environmental sample and magnetic shield sample containers, and capable of holding large surface samples; and the 7 1/2 X 8-inch Teflon documented sample bags in a 20-bag dispenser mounted on the lunar hand tool carrier. Both types of bags are stowed in the Apollo lunar sample return containers (ALSRC).

Tongs - The tongs are designed to allow the astronaut to retrieve small samples from the lunar surface while in a standing position. The tines are of such angles, length and number to allow samples of from 3/8 up to 2 1/2-inch diameter to be picked up. The tool is 32 inches long overall.

Spring scale - To weigh two rock boxes and other bags containing lunar material samples, to maintain weight budget for return to Earth.

Gnomon - This tool consists of a weighted staff suspended on a two-ring gimbal and supported by a tripod. The staff extends 12 inches above the gimbal and is painted with a gray scale and a color scale. The gnomon is used as a photographic reference to indicate local vertical, Sun angle and scale. The gnomon has a required accuracy of vertical indication of 20 minutes of arc. Magnetic damping is incorporated to reduce oscillations.

Color chart - The color chart is painted with three primary colors and a gray scale. It is used in calibration for lunar photography. The scale is mounted on the tool carrier but may easily be removed and returned to Earth for reference. The color chart is six inches in size.

Self-recording Penetrometer - Used in the soil mechanics experiment to measure the characteristics and mechanical properties of the lunar surface material. The penetrometer consists of a 30-inch penetration shaft and recording drum. Three interchangeable penetration cones (0.2, 0.5 and 1.0 square-inch cross sections) and a 1 X 5-inch pressure plate may be attached to the shaft. The crewman forces the penetrometer into the surface and a stylus scribes a force vs. depth plot on the recording drum. The drum can record up to 24 force-depth plots. The upper housing containing the recording drum is detached at the conclusion of the experiment for return to Earth and analysis by the principal investigator.

Lunar rake - Used by the crew for gathering samples ranging from one-half inch to one inch in size. The rake is adjustable and is fitted with stainless steel tines. A ten-inch rake handle adapts to the tool extension handle.

Apollo Lunar Hand Tool Carrier - An aluminum rack upon which the tools described above are stowed for lunar surface EVAs. The carrier differs from the folding carriers used on previous missions in that it mounts on the rear pallet of the lunar roving vehicle. The carrier may be hand carried during treks away from the LRV and is fitted with folding legs.

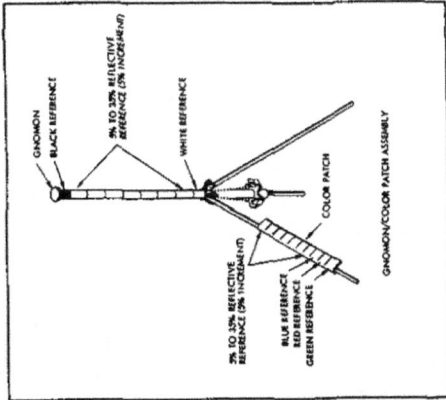

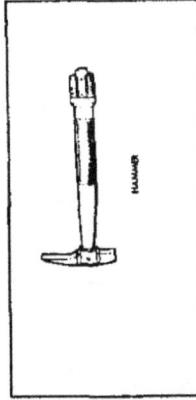

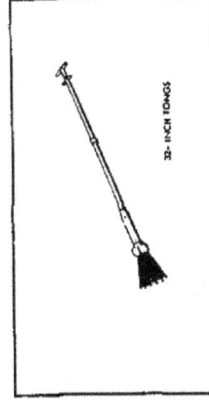

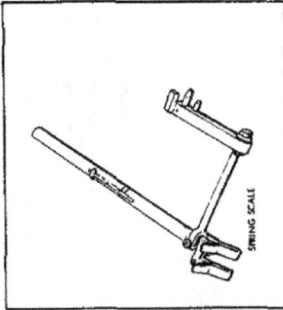

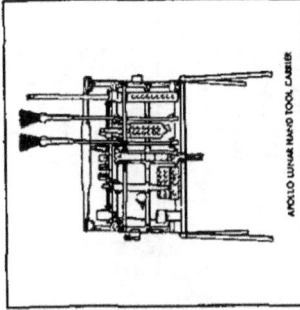

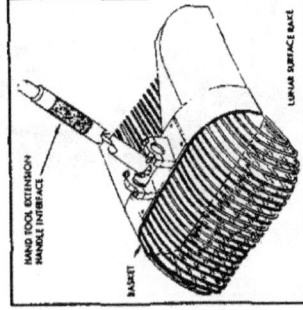

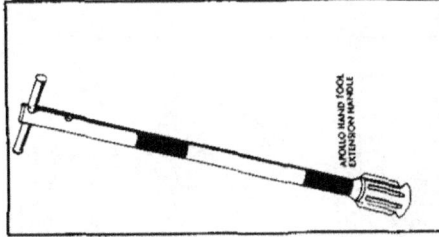

Lunar Geology Science Equipment

SPECIAL SAMPLE RETURN BAG

DOCUMENTED SAMPLE BAG

DIAGONAL SLIT IN TOP OF BAG

POCKETS

TEFLON HANDLES

SAMPLE COLLECTION BAG

TABS

NOTE: TEFLON BAGS ARE 7-1/2 IN. X 8 IN.

20-BAG DOCUMENTED SAMPLE BAG DISPENSER (2 PER ALSRC) ATTACHES TO ALHTC AND PLSS HARNESS

SPECIAL ENVIRONMENTAL SAMPLE CONTAINER

SAMPLE RETURN CONTAINER

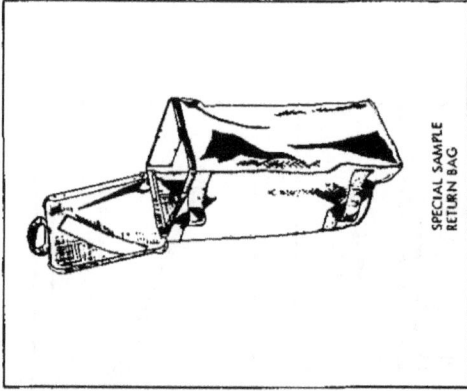

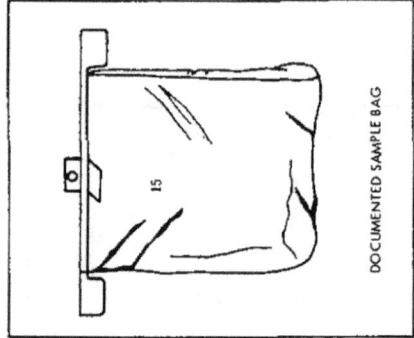

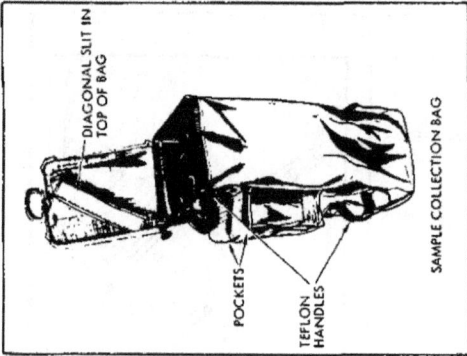

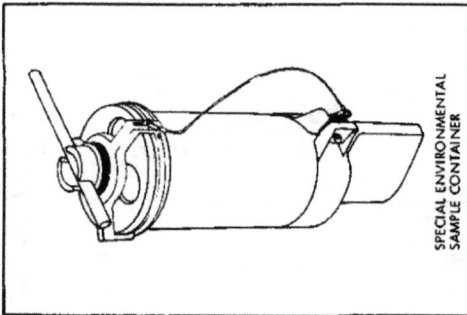

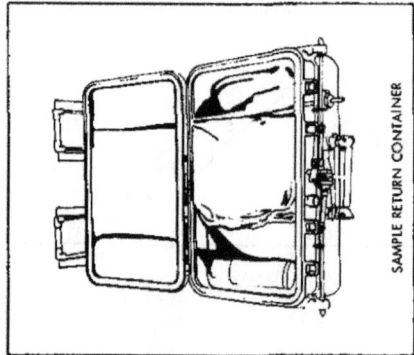

Lunar Geology Sample Containers

HADLEY-APENNINE LANDING SITE

The Apollo 15 landing site is located at 26° 04' 54"
North latitude by 3° 39' 30" East longitude at the foot of
the Apennine mountain range. The Apennines rise up to more
than 15,000 feet along the southeastern edge of the
Mare Imbrium (Sea of Rains).

The Apennine escarpment--highest on the Moon--is higher
above the flatlands than the east face of the Sierra Nevadas
in California and the Himalayan front rising above the plains
of India and Nepal. The landing site has been selected to
allow astronauts Scott and Irwin to drive from the LM to the
Apennine front during two of the EVAs.

A meandering canyon, Rima Hadley (Hadley Rille), approaches
the Apennine front near the landing site and the combination
of lurain provides an opportunity for the crew to explore
and sample a mare basin, a mountain front and a lunar rille in
a single mission.

Hadley Rille is a V-shaped gorge paralleling the Apennines
along the eastern edge of Mare Imbrium. The rille meanders
down from an elongated depression in the mountains and across
the Palus Putredenis (Swamp of Decay), merging with a second
rille about 62 miles (100 kilometers) to the north. Hadley
rille averages about a kilometer and a half in width and about
1,300 feet (400 meters) in depth throughout most of its length.

Large rocks have rolled down to the rille floor from
fresh exposures of what are thought to be stratified mare beds
along the tops of the rille walls. Selenographers are curious
about the origin of the Moon's sinuous rilles, and some
scientists believe the rilles were caused by some sort of fluid
flow mechanism--possibly volcanic.

Material sampled from the Apennines may yield specimens
of ancient rocks predating the formation and filling of the
major mare basins, while the rille may provide samples of
material dredged up by the impact of forming the 1.4-mile-wide
(2.2 km) Hadley C crater to the south of the landing site and on
the west side of Hadley rille. Secondary crater clusters in
the landing site vicinity are believed to have been formed by
ejecta from the Copernican-age craters Aristillus and Autolycus
which lie to the north of the landing site.

Mount Hadley, Hadley Rille and the various Hadley craters
in the region of the landing site are named for British scientist-
mathematician John Hadley (1682-1744) who made improvements in
reflector telescope design and invented the reflecting quadrant--
an ancestor of the mariner's sextant.

SITE SCIENCE RATIONALE

	APOLLO 11	APOLLO 12	APOLLO 14	APOLLO 15
TYPE	MARE	MARE	HILLY UPLAND	MOUNTAIN FRONT / RILLE / MARE
MATERIAL	BASALTIC LAVA	BASALTIC LAVA	DEEP-SEATED CRUSTAL MATERIAL	• DEEPER - SEATED CRUSTAL MATERIAL • BASALTIC LAVA
PROCESS	BASIN FILLING	BASIN FILLING	EJECTA BLANKET FORMATION	• MOUNTAIN SCARP • BASIN FILLING • RILLE FORMATION
AGE	OLDER MARE FILLING	YOUNGER MARE FILLING	EARLY HISTORY OF MOON o PRE-MARE MATERIAL o IMBRIUM BASIN FORMATION	• COMPOSITION AND AGE OF APENNINE FRONT MATERIAL • RILLE ORIGIN AND AGE • AGE OF IMBRIUM MARE FILL

4167

LUNAR ROVING VEHICLE

The lunar roving vehicle (LRV) will transport two astronauts on three exploration traverses of the Hadley-Apennine area of the Moon during the Apollo 15 mission. The LRV will also carry tools, scientific equipment, communications gear, and lunar samples.

The four-wheel, lightweight vehicle will greatly extend the lunar area that can be explored by man. The LRV can be operated by either astronaut.

The lunar roving vehicle will be the first manned surface transportation system designed to operate on the Moon. It marks the beginning of a new technology and represents an ambitious experiment to overcome many new and challenging problems for which there is no precedent in terrestrial vehicle design and operations.

First, the LRV must be folded up into a very small package in order to fit within the tight, pie-shaped confines of Quad 1 of the lunar module which will transport it to the Moon. After landing, the LRV must unfold itself from its stowed configuration and deploy itself to the lunar surface in its operational configuration, all with minimum assistance from the astronauts.

The lack of an atmosphere on the Moon, the extremes of surface temperature, the very small gravity, and the many unknowns associated with the lunar soil and topography impose requirements on the LRV which have no counterpart in Earth vehicles and for which no terrestrial experience exists. The fact that the LRV must be able to operate on a surface which can reach 250 degrees Fahrenheit and in a vacuum which rules out air cooling required the development of new concepts of thermal control.

The one-sixth gravity introduces a host of entirely new problems in vehicle dynamics, stability, and control. It makes much more uncertain such operations as turning, braking, and accelerating which will be totally different experiences than on Earth. The reduced gravity will also lead to large pitching, bouncing, and swaying motions as the vehicle travels over craters rocks, undulations, and other roughnesses of the lunar surface.

Many uncertainties also exist in the mechanical properties of the lunar soil involved in wheel/soil interaction. The interaction of lunar-soil mechanical properties, terrain roughness and vehicle controllability in one-sixth gravity will determine the performance of the LRV on the Moon.

Thus the LRV, while it is being used to increase the effectiveness of lunar exploration, will be exploring entirely new regimes of vehicle operational conditions in a new and hostile environment, markedly different from Earth conditions. The new knowledge to be gained from this mission should play an important role in shaping the course of future lunar and planetary exploration systems.

The LRV is built by the Boeing Co., Aerospace Group, at its Kent Space Center near Seattle, Wash., under contract to the NASA-Marshall Space Flight Center. Boeing's major subcontractor is the Delco Electronics Division of the General Motors Corp. Three flight vehicles have been built, plus seven test and training units, spare components, and related equipment.

General Description

The lunar roving vehicle is ten feet, two inches long; has a six-foot tread width; is 44.8 inches high; and has a 7.5-foot wheelbase. Each wheel is individually powered by a quarter-horsepower electric motor (providing a total of one horsepower) and the vehicle's top speed will be about eight miles an hour on a relatively smooth surface.

Two 36-volt batteries provide the vehicle's power, although either battery can power all vehicle systems if required. The front and rear wheels have separate steering systems, but if one steering system fails, it can be disconnected and the vehicle will operate with the other system.

Weighing approximately 460 pounds (Earth weight) when deployed on the Moon, the LRV will carry a total payload weight of about 1,080 pounds, more than twice its own weight. This cargo includes two astronauts and their portable life support systems (about 800 pounds), 100 pounds of communications equipment, 120 pounds of scientific equipment and photographic gear, and 60 pounds of lunar samples.

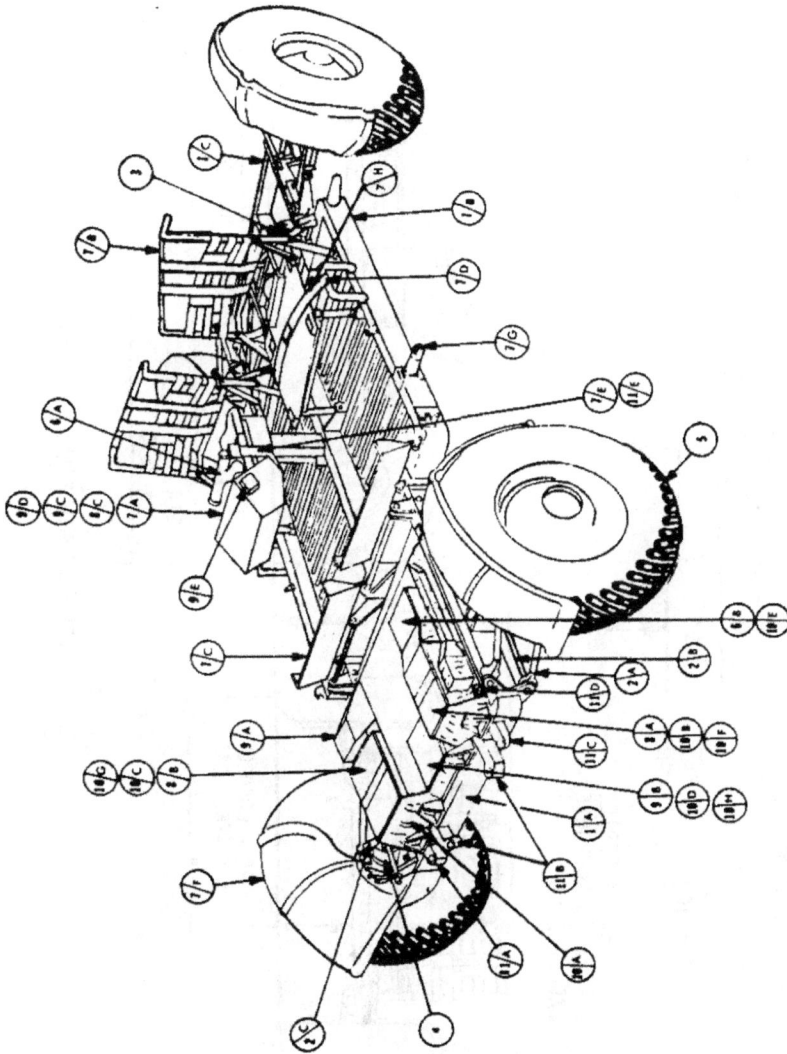

LRV WITHOUT STOWED PAYLOAD

① CHASSIS
A. FORWARD CHASSIS
B. CENTER CHASSIS
C. AFT CHASSIS

② SUSPENSION SYSTEM
A. SUSPENSION ARMS (UPPER AND LOWER)
B. TORSION BARS (UPPER AND LOWER)
C. DAMPER

③ STEERING SYSTEM (FORWARD AND AFT)

④ TRACTION DRIVE

⑤ WHEEL

⑥ DRIVE CONTROL
A. HAND CONTROLLER
B. DRIVE CONTROL ELECTRONICS (DCE)

⑦ CREW STATION
A. CONTROL AND DISPLAY CONSOLE
B. SEAT
C. FOOTREST
D. INBOARD HANDHOLD
E. OUTBOARD HANDHOLD
F. FENDER
G. TOEHOLD
H. SEAT BELT

⑧ POWER SYSTEM
A. BATTERY #1
B. BATTERY #2
C. INSTRUMENTATION

⑨ NAVIGATION
A. DIRECTIONAL GYRO UNIT (DGU)
B. SIGNAL PROCESSING UNIT (SPU)
C. INTEGRATED POSITION INDICATOR (IPI)
D. SUN SHADOW DEVICE
E. VEHICLE ATTITUDE INDICATOR

⑩ THERMAL CONTROL
A. INSULATION BLANKET
B. BATTERY NO. 1 DUST COVER
C. BATTERY NO. 2 DUST COVER
D. SPU DUST COVER
E. DCE THERMAL CONTROL UNIT
F. BATTERY NO. 1 RADIATOR
G. BATTERY NO. 2 RADIATOR
H. SPU THERMAL CONTROL UNIT

⑪ PAYLOAD INTERFACE
A. TV CAMERA RECEPTACLE
B. LCRU RECEPTACLE
C. HIGH GAIN ANTENNA RECEPTACLE
D. AUXILIARY CONNECTOR
E. LOW GAIN ANTENNA RECEPTACLE

AXIS REFERENCE

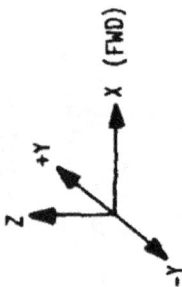

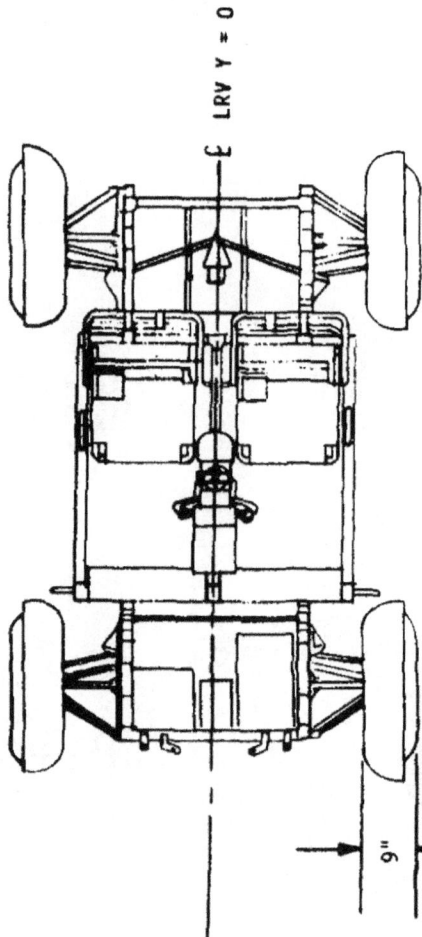

(DEPLOYED, EMPTY)
WEIGHT = 462 LB*

C.G. LOCATION:

X = 52.8
Y = -0.3
Z = 103.1

*INCLUDES BATTERIES
& PAYLOAD SUPPORTS,
EXCLUDES SSE.

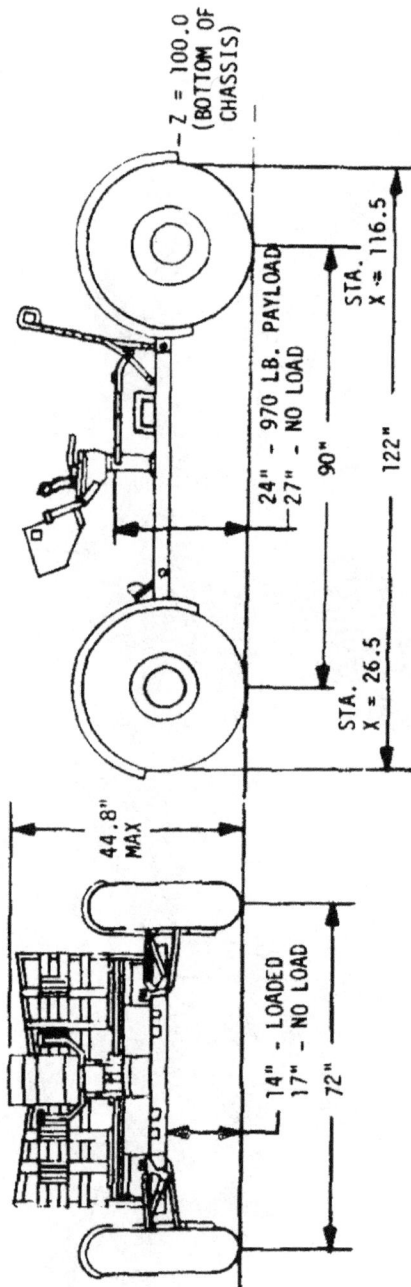

LRV COMPONENTS AND DIMENSIONS

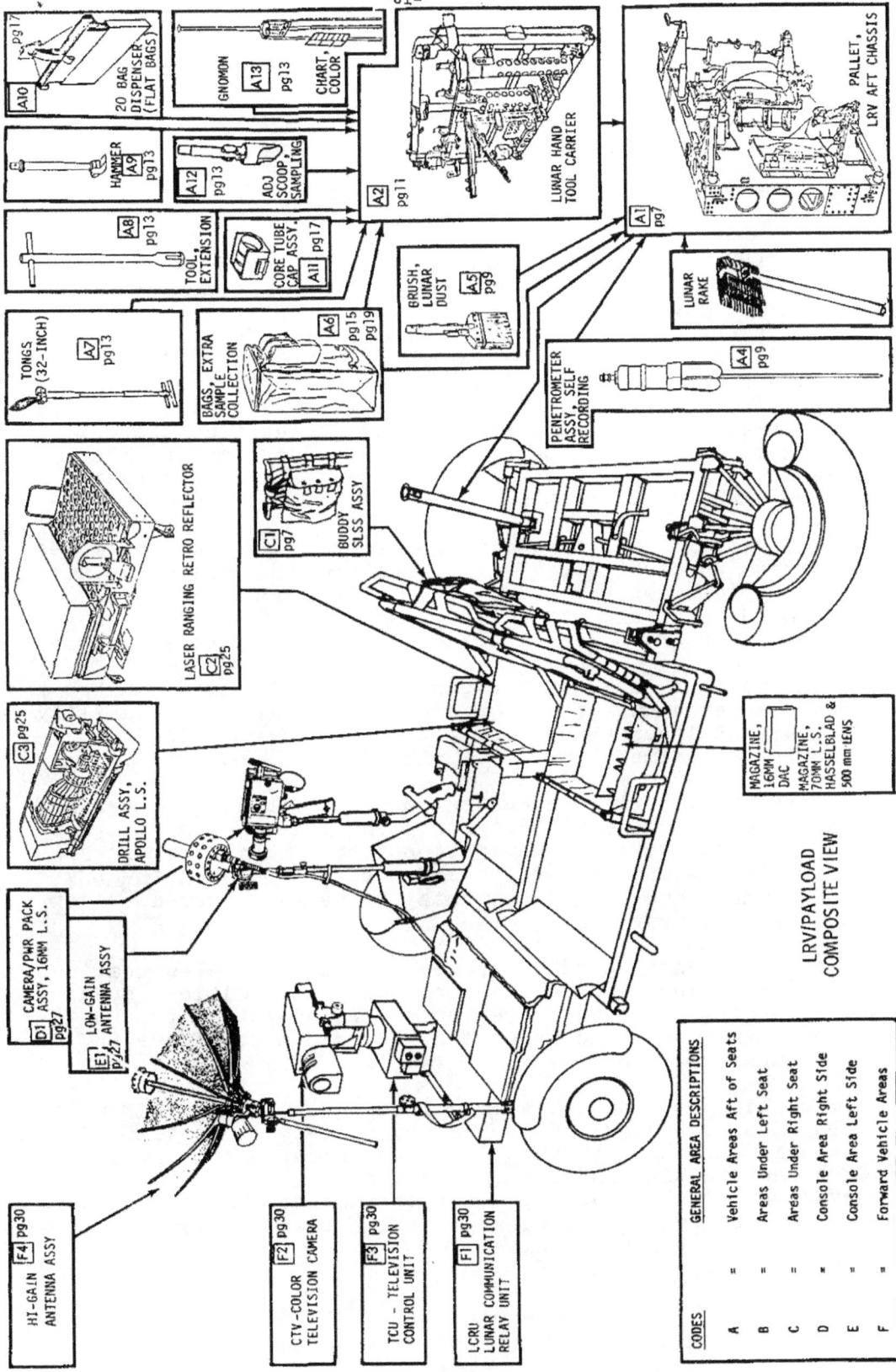

A10 pg17 — 20 BAG DISPENSER (FLAT BAGS)

A13 pg13 — GNOMON

CHART, COLOR

A9 pg13 — HAMMER

A12 pg13 — ADJ SCOOP, SAMPLING

A2 pg11 — LUNAR HAND TOOL CARRIER

A1 pg9 — PALLET, LRV AFT CHASSIS

A8 pg13 — TOOL, EXTENSION

A11 pg17 — CORE TUBE CAP ASSY,

A6 pg15 pg19 — BAGS, EXTRA SAMPLE COLLECTION

A5 pg9 — BRUSH, LUNAR DUST

LUNAR RAKE

A7 pg13 — TONGS (32-INCH)

A4 pg9 — PENTROMETER ASSY, SELF RECORDING

33

C2 pg25 — LASER RANGING RETRO REFLECTOR

C1 pg7 — BUDDY SLSS ASSY

C3 pg25 — DRILL ASSY, APOLLO L.S.

MAGAZINE, 16MM DAC

MAGAZINE, 70MM L.S. HASSELBLAD & 500 mm LENS

D1 pg27 — CAMERA/PWR PACK ASSY, 16MM L.S.

E1 pg27 — LOW-GAIN ANTENNA ASSY

LRV/PAYLOAD COMPOSITE VIEW

F4 pg30 — HI-GAIN ANTENNA ASSY

F2 pg30 — CTV-COLOR TELEVISION CAMERA

F3 pg30 — TCU - TELEVISION CONTROL UNIT

F1 pg30 — LCRU LUNAR COMMUNICATION RELAY UNIT

CODES		GENERAL AREA DESCRIPTIONS
A	=	Vehicle Areas Aft of Seats
B	=	Areas Under Left Seat
C	=	Areas Under Right Seat
D	=	Console Area Right Side
E	=	Console Area Left Side
F	=	Forward Vehicle Areas

The LRV will travel to the Moon folded inside stowage Quadrant 1 of the lunar module's descent stage. During the first lunar surface EVA period the astronauts will manually deploy the vehicle and prepare it for cargo loading and operation.

The LRV is designed to operate for 78 hours during the lunar day. It can make several exploration sorties up to a cumulative distance of 40 miles (65 kilometers). Because of limitations in the astronauts' portable life support systems (PLSS), however, the vehicle's range will be restricted to a radius of about six miles (9.5 kilometers) from the lunar module. This provides a walk-back capability to the LM should the LRV become immobile at the maximum radius from the LM. This six-mile radius contains about 113 square miles which is available for investigation, some ten times the area that could be explored on foot.

The vehicle is designed to negotiate step-like obstacles 9.8 inches (25 centimeters) high, and cross crevasses 22.4 inches (50 centimeters) wide. The fully loaded vehicle can climb and descend slopes as steep as 25 degrees. A parking brake can stop and hold the LRV on slopes of up to 30 degrees.

The vehicle has ground clearance of at least 14 inches (35 centimeters) on a flat surface. Pitch and roll stability angles are at least 45 degrees with a full load. The turn radius is approximately 10 feet with forward and aft steering.

Both crewmen will be seated so that both front wheels are visible during normal driving. The driver will navigate through a dead reckoning navigation system that determines the vehicle heading, direction and distance between the LRV and the lunar module, and the total distance traveled at any point during a traverse.

The LRV has five major systems: mobility, crew station, navigation, power, and thermal control. In addition, space support equipment includes mechanisms which attach the LRV to the lunar module and which enable deployment of the LRV to the lunar surface.

Auxiliary equipment (also called stowed equipment) will be provided to the LRV by the Manned Spacecraft Center, Houston. This equipment includes the lunar communications relay unit (LCRU) and its high and low gain antennas, the ground control television assembly (GCTA), a motion picture camera, scientific equipment, astronaut tools, and sample stowage bags.

Mobility System

The mobility system has the largest number of subsystems, including the chassis, wheels, traction drive, suspension, steering, and drive control electronics.

The aluminum chassis is divided into forward, center and aft sections that support all equipment and systems. The forward section holds both batteries, the navigation system's signal processing unit and directional gyroscope, and the drive control electronics (DCE) for traction drive and steering.

The center section holds the crew station, with its two seats, control and display console, and hand controller. This section's floor is made of beaded aluminum panels, structurally capable of supporting the full weight of both astronauts standing in lunar gravity. The aft section is a platform for the LRV's scientific payload. The forward and aft chassis sections fold over the center section and lock in place during stowage in the lunar module.

Each LRV wheel has a spun aluminum hub and a titanium bump stop (inner frame) inside the tire (outer frame). The tire is made of a woven mesh of zinc-coated piano wire to which titanium treads are riveted in a chevron pattern around the outer circumference. The bump stop prevents excess deflection of the outer wire mesh during heavy impact. Each wheel weighs 12 pounds on Earth (two lunar pounds) and is designed for a driving distance of at least 112 statute miles (180 kilometers). The wheels are 32 inches in diameter and nine inches wide.

The traction drive attached to each wheel consists of a harmonic drive unit, a drive motor, and a brake assembly. The harmonic drives reduce motor speed at the rate of 80-to-1, allowing continuous vehicle operation at all speeds without gear shifting. Each drive has an odometer pickup that transmits magnetic pulses to the navigation system's signal processing unit. (Odometers measure distance travelled.)

The quarter-horsepower, direct current, brush-type drive motors normally operate from a 36-volt input. Motor speed control is furnished from the drive control electronics package. Suspension system fittings on each motor form the king-pin for the vehicle's steering system.

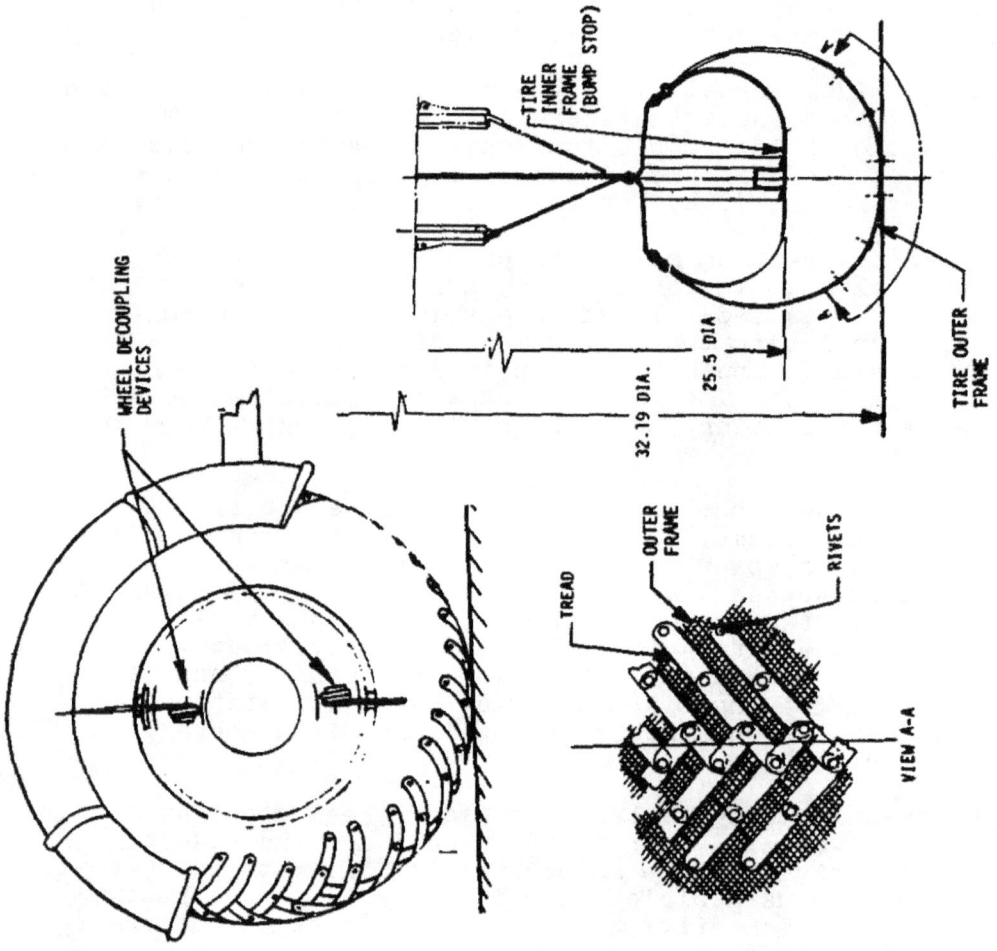

WHEEL DECOUPLING DEVICES

TIRE INNER FRAME (BUMP STOP)

32.19 DIA.

25.5 DIA

TIRE OUTER FRAME

LRV WHEEL

TREAD

OUTER FRAME

RIVETS

VIEW A-A

The traction drive is equipped with a mechanical brake, cable-connected to the hand controller. Moving the controller rearward de-energizes the drive motor and forces hinged brake shoes against a brake drum, stopping rotation of the wheel hub about the harmonic drive. Full rearward movement of the controller engages and locks the parking brakes. To disengage the parking brake, the controller is moved to the steer left position at which time the brake releases and the controller is allowed to return to neutral.

Each wheel can be manually uncoupled from the traction drive and brake to allow "free-wheeling" about the drive housing, independent of the drive train. The same mechanism will re-engage a wheel.

The chassis is suspended from each wheel by a pair of parallel arms mounted on torsion bars and connected to each traction drive. A damper (shock absorber) is a part of each suspension system. Deflection of the system and the tires allows a 14-inch ground clearance when the vehicle is fully loaded, and 17 inches when unloaded. The suspension systems can be folded about 135 degrees over the center chassis for stowage in the lunar module.

Both the front and rear wheels have independent steering systems that allow a "wall-to-wall" turning radius of 122 inches (exactly the vehicle's length). Each system has a small, 1/10th-horsepower, 5,000-rpm motor driving through a 257-to-1 reduction into a gear that connects with the traction drive motor by steering arms and a tie rod. A steering vane, attached between the chassis and the steering arms, allows the extreme steering angles required for the short turn radius.

If a steering malfunction occurs on either the front or rear steering assembly, the steering linkage to that set of wheels can be disengaged and the mission can continue with the remaining active steering assembly. A crewman can reconnect the rear steering assembly if desired.

The vehicle is driven by a T-shaped hand controller located on the control and display console post between the two crewmen. The controller maneuvers the vehicle forward, reverse, left and right, and controls speed and braking.

- more -

A knob that determines whether the vehicle moves forward or reverse is located on the T-handle's vertical stem. With the knob pushed down, the hand controller can only be moved forward. When the knob is pushed up and the controller moved rearward, the LRV can be operated in reverse.

Drive control electronics accept forward and reverse speed control signals from the hand controller, and electronic circuitry will switch drive power off and on. The electronics also provide magnetic pulses from the wheels to the navigation system for odometer and speedometer readouts.

Crew Station

The LRV crew station consists of the control and display console, seats and seat belts, an armrest, footrests, inboard and outboard handholds, toeholds, floor panels, and fenders.

The control and display console is separated into two main parts: the upper portion holds navigation system gauges and the lower portion holds vehicle monitors and controls.

Attached to the upper left side of the console is an attitude indicator that shows vehicle pitch and roll. Pitch is indicated upslope or downslope within a range of + 25 degrees; roll is indicated as 25 degrees left or right. Readings, normally made with the vehicle stopped, are transmitted verbally to Houston's Mission Control Center for periodic navigation computation.

At the console's top left is an integrated position indicator (IPI). The indicator's outer circumference is a large dial that shows the vehicle's heading (direction) with respect to lunar north. Inside the circular dial are three indicators that display readings of bearing, distance and range. The bearing indicator shows direction to the Lunar Module, the distance indicator records distance traveled by the LRV, and the range indicator displays distance to the Lunar Module.

The distance and range indicators have total scale capacities of 99.9 kilometers (62 statute miles). If the navigation system loses power, the bearing and range readings will remain displayed.

LRV CREW STATION COMPONENTS - CONTROL AND DISPLAY CONSOLE

HAND CONTROLLER OPERATION:

T-HANDLE PIVOT FORWARD - INCREASED DEFLECTION FROM NEUTRAL INCREASES FORWARD SPEED.

T-HANDLE PIVOT REARWARD - INCREASED DEFLECTION FROM NEUTRAL INCREASES REVERSE SPEED.

T-HANDLE PIVOT LEFT - INCREASED DEFLECTION FROM NEUTRAL INCREASES LEFT STEERING ANGLE.

T-HANDLE PIVOT RIGHT - INCREASED DEFLECTION FROM NEUTRAL INCREASES RIGHT STEERING ANGLE.

T-HANDLE DISPLACED REARWARD - REARWARD MOVEMENT INCREASES BRAKING FORCE. FULL 3 INCH REARWARD APPLIES PARKING BRAKE. MOVING INTO BRAKE POSITION DISABLES THROTTLE CONTROL AT 15° MOVEMENT REARWARD.

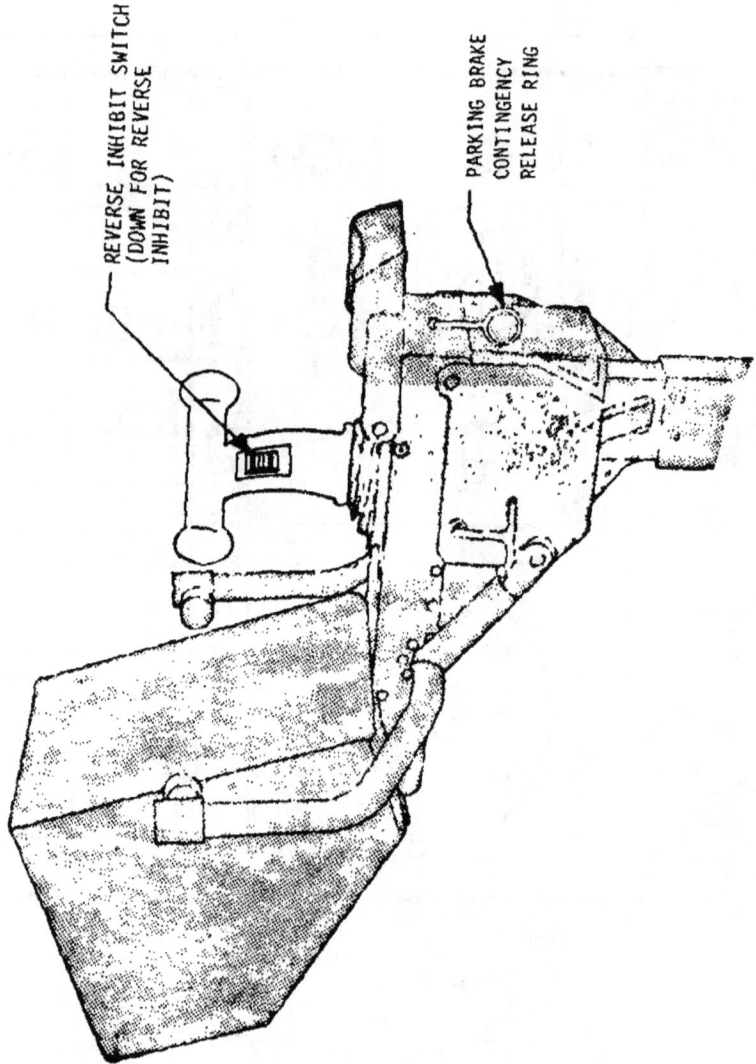

REVERSE INHIBIT SWITCH (DOWN FOR REVERSE INHIBIT)

PARKING BRAKE CONTINGENCY RELEASE RING

HAND CONTROLLER

In the center of the console's upper half is a Sun shadow device (Sun compass) that can determine the LRV's heading with respect to the Sun. The device casts a shadow on a graduated scale when it is pulled up at right angles from the console. The point where the Sun's shadow intersects the scale will be read by the crew to Mission Control, which will tell the crew what heading to set into the navigation system. The device can be used at Sun elevation angles up to 75 degrees.

A speed indicator shows LRV velocity from 0 to 20 kilometers an hour (0-12 statute mph). This display is driven by odometer pulses from the right rear wheel through the navigation system's signal processing unit.

A gyro torquing switch adjusts the heading indicator during navigation system resettings, and a system reset switch returns the bearing, distance, and range indicators to zero.

Down the left side of the console's lower half are switches that allow power from either battery to feed a dual bus system. Next to these switches are two power monitors that give readings of ampere hours remaining in the batteries, and either volts or amperes from each battery. To the right of these are two temperature monitors that show readouts from the batteries and the drive motors. Below these monitors are switches that control the steering motors and drive motors.

An alarm indicator (caution and warning flag) atop the console pops up if a temperature goes above limits in either battery or in any of the drive motors. The indicator can be reset.

The LRV's seats are tubular aluminum frames spanned by nylon strips. They are folded flat onto the center chassis during launch and are erected by the crewmen after the LRV is deployed. The seat backs support and restrain the astronauts' portable life support systems (PLSS) from moving sideways when crewmen are sitting on the LRV. The seat bottoms have cutouts for access to PLSS flow control valves and provisions for vertical support of the PLSS. The seat belts are made of nylon webbing. They consist of an adjustable web section and a metal hook that is snapped over the outboard handhold.

- more -

The armrest, located directly behind the hand controller, supports the arm of the crewman who is using the controller. The footrests are attached to the center floor section and may be adjusted prior to launch if required to fit each crewman. They are stowed against the center chassis floor and secured by pads until deployment by the crewmen.

The inboard handholds are made of one-inch aluminum tubing and help the crewmen get in and out of the LRV. The handholds also have attachment receptacles for the 16mm camera and the low gain antenna (auxiliary equipment). The outboard handholds are integral parts of the chassis and provide crew comfort and stability when seated on the LRV.

Toeholds are provided to help crewmen leave the LRV. They are made by dismantling the LRV support tripods and inserting the tripod center member legs into chassis receptacles on each side of the vehicle to form the toeholds. The toeholds also can be used as a tool to engage and disengage the wheel decoupling mechanism.

The vehicle's fenders, made of lightweight fiberglass, are designed to prevent lunar dust from being thrown on the astronauts, their scientific payload, and sensitive vehicle parts, or from obstructing astronaut vision while driving. The fender front and rear sections are retracted during flight and extended by the crewmen after deployment.

Navigation System

The dead reckoning navigation system is based on the principle of starting a sortie from a known point, recording direction relative to the LM and distance traveled, and periodically calculating vehicle position relative to the LM from these data.

The system contains three major components: a directional gyroscope that provides the vehicle's heading; odometers on each wheel's traction drive unit that provide distance information; and a signal processing unit (essentially a small, solid-state computer) that determines bearing and range to the LM, distance traveled, and velocity.

All navigation system readings are displayed on the control and display console. Components are activated by pressing the system reset button, which moves all digital displays and internal registers to zero. The system will be reset at the beginning of each LRV traverse.

The directional gyroscope is aligned by measuring the inclination of the LRV (using the attitude indicator) and measuring vehicle orientation with respect to the Sun (using the Sun shadow device). This information is relayed to Mission Control, where a heading angle is calculated and read back to the crew. The gyro is then adjusted until the heading indicator reads the same as the calculated value.

Nine odometer magnetic pulses are generated for each wheel revolution, and these signals enter logic in the signal processing unit (SPU). The SPU selects pulses from the third fastest wheel to insure that the pulses are not based on a wheel that has inoperative odometer pulses or has excessive slip. (Because the SPU cannot distinguish between forward and reverse wheel rotation, reverse operation of the vehicle will add to the odometer reading.) The SPU sends outputs directly to the distance indicator and to the range and bearing indicators through its digital computer. Odometer pulses from the right rear wheel are sent to the speed indicator.

The Sun shadow device is a kind of compass that can determine the LRV's heading in relation to the Sun. It will be used at the beginning of each sortie to establish the initial heading, and then be used periodically during sorties to check for slight drift in the gyro unit.

Power System

The power system consists of two 36-volt, non-rechargeable batteries, distribution wiring, connectors, switches, circuit breakers, and meters to control and monitor electrical power.

The batteries are encased in magnesium and are of plexiglass monoblock (common cell walls) construction, with silver-zinc plates in potassium hydroxide electrolyte. Each battery has 23 cells and a 115-ampere-hour capacity.

Both batteries are used simultaneously with an approximately equal load during LRV operation. Each battery can carry the entire LRV electrical load, however, and the circuitry is designed so that, if one battery fails, the load can be switched to the other battery.

The batteries are located on the forward chassis section, enclosed by a thermal blanket and dust covers. Battery No. 1 (left side) is connected thermally to the navigation system's signal processing unit and is a partial heat sink for that unit. Battery No. 2 (right side) is thermally tied to the navigation system's directional gyro and serves it as a heat sink.

The batteries are activated when installed on the LRV at the launch pad about five days before launch. They are monitored for voltage and temperature on the ground until about T-20.5 hours in the countdown. On the Moon the batteries are monitored for temperature, voltage, output current, and remaining ampere hours through displays on the control console.

During normal LRV operation, all mobility power will be turned off if a stop is to exceed five minutes, but the navigation system's power will stay on during each complete sortie.

For battery survival their temperature must remain between 40 and 125 degrees F. When either battery reaches 125 degrees, or when any motor reaches 400 degrees, temperature switches actuate to flip up the caution and warning flag atop the control console.

An auxiliary connector, located at the front of the vehicle, provides 150 watts of 36-volt power for the lunar communications relay unit (LCRU), whose power cable is attached to the connector before launch.

Thermal Control

Thermal control is used on the LRV to protect temperature-sensitive components during all phases of the mission. Thermal controls include special surface finishes, multi-layer insulation, space radiators, second-surface mirrors, thermal straps, and fusible mass heat sinks.

The basic concept of thermal control is to store heat while the vehicle is running and to cool by radiation between sorties.

During operation, heat is stored in several thermal fusible mass tank heat sinks and in the two batteries. Space radiators are located atop the signal processing unit, the drive control electronics, and the batteries. Fused silica second-surface mirrors are bonded to the radiators to lessen solar energy absorbed by the exposed radiators. The radiators are only exposed while the LRV is parked between sorties.

During sorties, the radiators are protected from lunar surface dust by three dust covers. The radiators are manually opened at the end of each sortie and held by a latch that holds them open until battery temperatures cool down to 45 degrees F (+5 degrees), at which time the covers automatically close.

A multi-layer insulation blanket protects components from harsh environments. The blanket's exterior and some parts of its interior are covered with a layer of Beta cloth to protect against wear.

All instruments on the control and display console are mounted to an aluminum plate isolated by radiation shields and fiberglass mounts. Console external surfaces are coated with thermal control paint and the face plate is anodized, as are all handholds, footrests, seat tubular sections, and center and aft floor panels.

Stowage and Deployment

Certain LRV equipment (called space support equipment) is required to attach the folded vehicle to the Lunar Module during transit to the Moon and during deployment on the surface.

The LRV's forward and aft chassis sections, and the four suspension systems, are folded inward over the center chassis inside the LM's Quadrant 1. The center chassis' aft end is pointing up, and the LRV is attached to the LM at three points.

The upper point is attached to the aft end of the center chassis and the LM through a strut that extends horizontally from the LM quadrant's apex. The lower points are attached between the forward sides of the center chassis, through the LRV tripods, to supports in the LM quadrant.

LRV DEPLOYMENT SEQUENCE

- AFT CHASSIS UNFOLDS
- REAR WHEELS UNFOLD
- AFT CHASSIS LOCKS IN POSITION

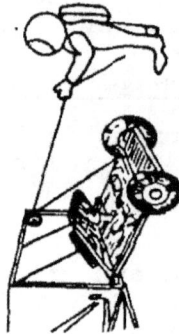

- ASTRONAUT UNFOLDS SEATS, FOOTRESTS, ETC. (FINAL STOP)

- ASTRONAUT LOWERS LRV FROM STORAGE BAY WITH FIRST REEL

- FORWARD CHASSIS LOCKS IN POSITION. ASTRONAUT LOWERS LRV TO SURFACE WITH SECOND REEL.

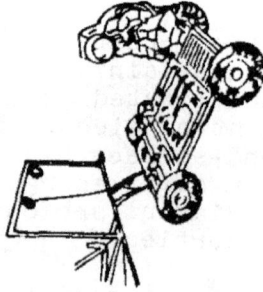

- LRV STOWED IN QUADRANT
- ASTRONAUT REMOTELY INITIATES AND EXECUTES DEPLOYMENT

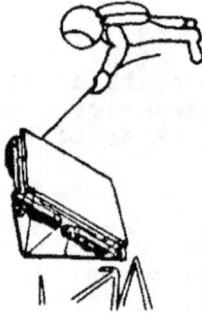

- FORWARD CHASSIS UNFOLDS
- FRONT WHEELS UNFOLD

The vehicle's deployment mechanism consists of the cables, shock absorbers, pin retract mechanisms, telescoping tubes, pushoff rod, and other gear.

LRV deployment is essentially manual. A crewman first releases a mylar deployment cable, attached to the center rear edge of the LRV's aft chassis. He hands the cable to the second crewman who stands by during the entire deployment operation ready to help the first crewman.

The first crewman then ascends the LM ladder part-way and pulls a D-ring on the side of the descent stage. This deploys the LRV out at the top about five inches (4 degrees) until it is stopped by two steel deployment cables, attached to the upper corners of the vehicle. The crewman then descends the ladder, walks around to the LRV's right side and pulls the end of a mylar deployment tape from a stowage bag in the LM. The crewman unreels this tape, hand-over-hand, to deploy the vehicle.

As the tape is pulled, two support cables are unreeled, causing a pushoff tube to push the vehicle's center of gravity over-center so it will swivel outward from the top. When the chassis reaches a 45-degree angle from the LM, release pins on the forward and aft chassis are pulled, the aft chassis unfolds, the aft wheels are unfolded by the upper torsion bars and deployed, and all latches are engaged.

As the crewman unwinds the tape, the LRV continues lowering to the surface. At a 73-degree angle from the LM, the forward chassis and wheels are sprung open and into place. The crewman continues to pull the deployment tape until the aft wheels are on the surface and the support cables are slack. He then removes the two slack cables from the LRV and walks around the vehicle to its left side. There he unstows a second mylar deployment tape. Pulling this tape completes the lowering of the vehicle to the surface and causes telescoping tubes attached between the LM and the LRV's forward end to guide the vehicle away from the LM. The crewman then pulls a release lanyard on the forward chassis' right side that allows the telescoping tubes to fall away.

The two crewmen then deploy the fender extensions on each wheel, insert the toeholds, deploy the handholds and footrests, set the control and display console in its upright position, release the seat belts, unfold the seats, and remove locking pins and latches from several places on the vehicle.

One crewman will then board the LRV and make sure
that all controls are working. He will back the vehicle
away from the Lunar Module and drive it to a position
near the LM quadrant where the auxiliary equipment is
stored, verifying as he drives that all LRV controls and
displays are operating. At the new parking spot, the
LRV will be powered down while the two astronauts load
the auxiliary equipment aboard the vehicle.

Development Background

The manned Lunar Roving Vehicle development program
began in October 1969 when the Boeing Co. was awarded a
contract to build four (later changed to three) flight
model LRVs. The Apollo 15 LRV was delivered to NASA
March 15, 1971, two weeks ahead of schedule, and less
than 17 months after contract award.

During this extremely short development and test
program, more than 70 major tests have been conducted by
Boeing and its major subcontractor, GM's Delco Electronics
Division. Tests and technical reviews have been held at
Boeing's Kent, Washington plant; at Delco's laboratories
near Santa Barbara, California, and at NASA's Marshall
Space Flight Center, Manned Spacecraft Center, and Kennedy
Space Center.

Seven LRV test units have been built to aid development
of the three flight vehicles: an LRV mass unit to determine
if the LRV's weight might cause stresses or strains in the
LM's structure; two one-sixth-weight units to test the LRVs
deployment mechanism; a mobility unit to test the mobility
system which was later converted to an Earth trainer (one-G
trainer) unit for astronaut training; a vibration unit to
verify the strength of the LRV structure; and a qualification
unit to test vibrations, temperature extremes and vacuums to
prove that the LRV will withstand all operating conditions.

Boeing produces the vehicle's chassis, crew station,
navigation system, power system, deployment system, ground
support equipment, and vehicle integration and assembly.
Delco produces the mobility system and built the one-G
astronaut training vehicle. Eagle-Picher Industries, Inc.,
Joplin, Missouri builds the LRV batteries, and the United
Shoe Machinery Corp., Wakefield, Massachusetts provides the
harmonic drive unit.

LUNAR COMMUNICATIONS RELAY UNIT (LCRU)

The range from which the Apollo 15 crew can operate from the lunar module during EVAs is extended over the lunar horizon by a suitcase-size device called the lunar communications relay unit (LCRU). The LCRU acts as a portable relay station for voice, TV, and telemetry directly between the crew and Mission Control Center instead of through the lunar module communications system.

Completely self-contained with its own power supply and folding S-Band antenna, the LCRU may be mounted on a rack at the front of the lunar roving vehicle (LRV) or hand-carried by a crewman. In addition to providing communications relay, the LCRU relays ground-command signals to the ground commanded television assembly (GCTA) for remote aiming and focussing the lunar surface color television camera. The GCTA is described in another section of this press kit.

Between stops with the lunar roving vehicle, crew voice is beamed Earthward by a low-gain helical S-Band antenna. At each traverse stop, the crew must boresight the high-gain parabolic antenna toward Earth before television signals can be transmitted. VHF signals from the crew portable life support system (PLSS) transceivers are converted to S-Band by the LCRU for relay to the ground, and conversely, from S-Band to VHF on the uplink to the EVA crewmen.

The LCRU measures 22 x 16 x 6 inches not including antennas, and weighs 55 Earth pounds (9.2 lunar pounds). A protective thermal blanket around the LCRU can be peeled back to vary the amount of radiation surface which consists of 196 square inches of radiating mirrors to reflect solar heat. Additionally, wax packages on top of the LCRU enclosure stabilize the LCRU temperature by a melt-freeze cycle. The LCRU interior is pressurized to 7.5 psia differential (one-half atmosphere).

Internal power is provided to the LCRU by a 19-cell silver-zinc battery with a potassium hydroxide electrolyte. The battery weighs nine Earth pounds (1.5 lunar pounds) and measures 4.7 x 9.4 x 4.65 inches. The battery is rated at 400 watt hours, and delivers 29 volts at a 3.1-ampere current load. The LCRU may also draw power from the LRV batteries.

Three types of antennas are fitted to the LCRU system: a low-gain helical antenna for relaying voice and data when the LRV is moving and in other instances when the high-gain antenna is not deployed; a three-foot diameter parabolic rib-mesh high-gain antenna for relaying a television signal; and a VHF omni antenna for receiving crew voice and data from the PLSS transceivers. The high-gain antenna has an optical sight which allows the crewman to boresight on Earth for optimum signal strength. The Earth subtends one-half degree angle when viewed from the lunar surface.

The LCRU can operate in several modes: mobile on the LRV, fixed base such as when the LRV is parked, hand-carried in contingency situations such as LRV failure, and remote by ground control for tilting the television camera to picture LM ascent.

Detailed technical and performance data on the LCRU is available at the Houston News Center query desk.

TELEVISION AND

GROUND CONTROLLED TELEVISION ASSEMBLY

Two different color television cameras will be used during the Apollo 15 mission. One, manufactured by Westinghouse, will be used in the command module. It will be fitted with a two-inch black and white monitor to aid the crew in focus and exposure adjustment.

The other camera, manufactured by RCA, is for lunar surface use and will be operated from three different positions-mounted on the LM MESA, mounted on a tripod and connected to the LM by a 100-foot cable, and installed in the LRV with signal transmission through the lunar communication relay unit rather than through the LM communications system as in the other two models.

While on the LRV, the camera will be mounted on the ground controlled television assembly (GCTA). The camera can be aimed and controlled by astronauts or it can be remotely controlled by personnel located in the Mission Control Center. Remote command capability includes camera "on" and "off", pan, tilt, zoom, iris open/closed (f2.2 to f22) and peak or average automatic light control.

The GCTA is capable of tilting the TV camera upward 85 degrees, downward 45 degrees, and panning the camera 340 degrees between mechanical stops. Pan and tilt rates are three degrees per second.

The TV lens can be zoomed from a focal length of 12.5mm to 75mm corresponding to a field of view from three to nine degrees.

At the end of the third EVA, the crew will park the LRV about 300 feet east of the LM so that the color TV camera can cover the LM ascent from the lunar surface. Because of a time delay in a signal going the quarter million miles out to the Moon, Mission Control must anticipate ascent engine ignition by about two seconds with the tilt command.

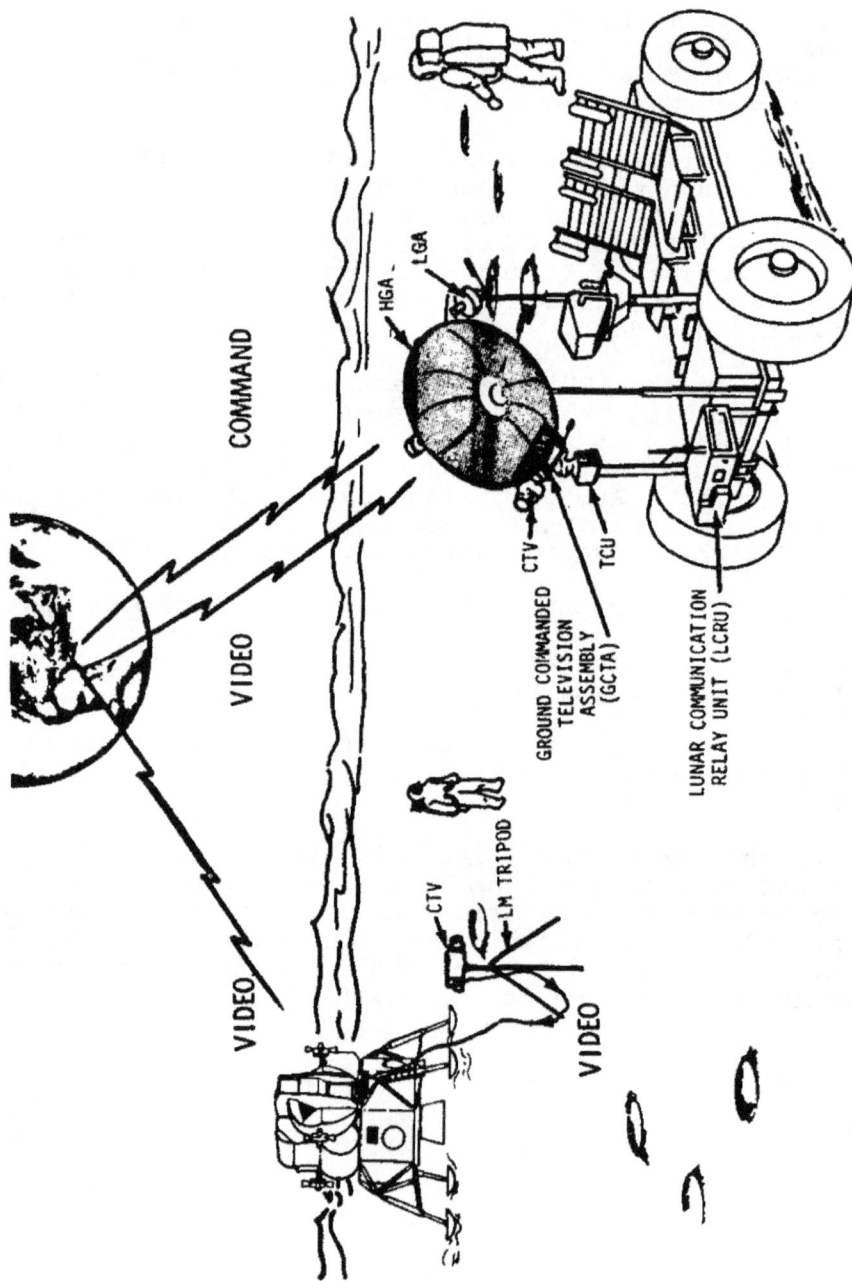

GCTA SYSTEM

COMMAND

VIDEO

VIDEO

VIDEO

VIDEO

HGA

LGA

CTV

TCU

GROUND COMMANDED
TELEVISION
ASSEMBLY
(GCTA)

LUNAR COMMUNICATION
RELAY UNIT (LCRU)

CTV

LM TRIPOD

It is planned to view the solar eclipse occurring
on August 6, if sufficient battery power remains. The total
eclipse extends from 2:24 p.m. EDT to 5:06 p.m. EDT, bracketed
by periods of partial eclipse. During the solar eclipse,
the camera will be used to make several lunar surface, solar
and astronomical observations. Of particular importance
will be the observations of the lunar surface under changing
lighting conditions. Observations planned during this
period include views of the LM, the crescent Sun, the corona
edge, the Apennine front, the zodiacal light, the Milky
Way, Saturn, Mercury, the eclipse ring, foreground rocks,
the lunar horizon and other lunar surface features.

The GCTA and camera each weigh approximately 13 pounds.
The overall length of the camera is 18.1 inches, its width
is 6.7 inches, and its height is 10.13 inches. The GCTA,
and LCRU are built by RCA.

PROBABLE AREAS FOR NEAR LM LUNAR SURFACE ACTIVITIES

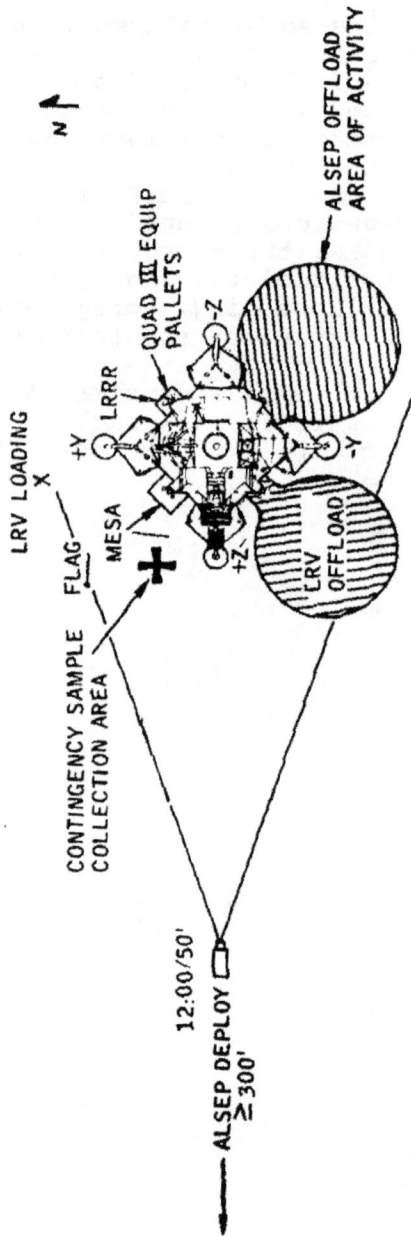

APOLLO 15 TV SCHEDULE

Day	Date	CDT	GET	Duration	Activity	Vehicle	Station
Monday	July 26	11:59 am	3:25	25 min.	Transposition and docking	CSM	Goldstone
Tuesday	July 27	6:20 pm	33:46	45 min.	IVA to LM	CSM	Goldstone
Friday	July 30	9:22 am	96:48	14 min	Landing Site out window	SCM	Madrid
Saturday	July 31	8:34 am	120:00	6 hr. 40 min.	EVA 1	LM/LRV	Honeysuckle/ Madrid
Sunday	Aug. 1	6:09 am	141:35	6 hr. 25 min.	EVA 2	LM/LRV	Parkes/ Honeysuckle/ Madrid
Monday	Aug. 2	2:49 am	162:15	5 hr. 45 min.	EVA 3	LM/LRV	Parkes/ Honeysuckle/ Madrid
Monday	Aug. 2	12:04 am	171:30	30 min.	LM liftoff	LRV	Madrid
Monday	Aug. 2	1:37 pm	173:03	6 min.	Rendezvous	CSM	Madrid
Monday	Aug. 2	2:00 pm	173:26	5 min.	Docking	CSM	Madrid
Thursday	Aug. 5	10:44 am	242:10	30 min.	Trans-Earth EVA	CSM	Honeysuckle
Friday	Aug. 6	3:00 pm	270:26	30 min.	Trans-Earth Coast	CSM	Madrid

PHOTOGRAPHIC EQUIPMENT

Still and motion pictures will be made of most space-craft maneuvers and crew lunar surface activities. During lunar surface operations, emphasis will be on documenting placement of lunar surface experiments and on recording in their natural state the lunar surface features.

Command Module lunar orbit photographic tasks and experiments include high-resolution photography to support future landing missions, photography of surface features of special scientific interest and astronomical phenomena such as Gegenschein, zodiacal light, libration points, galactic poles and the Earth's dark side.

Camera equipment stowed in the Apollo 15 command module consists of one 70mm Hasselblad electric camera, a 16mm Maurer motion picture camera, and a 35mm Nikon F single-lens reflex camera. The command module Hasselblad electric camera is normally fitted with an 80mm f/2.8 Zeiss Planar lens, but a bayonet-mount 250mm lens can be fitted for long-distance Earth/Moon photos. A 105mm f/4.3 Zeiss UV Sonnar is provided for the ultraviolet photography experiment.

The 35mm Nikon F is fitted with a 55mm f/1.2 lens for the Gegenschein and dim-light photographic experiments.

The Maurer 16mm motion picture camera in the command module has lenses of 10, 18 and 75mm focal length available. Accessories include a right-angle mirror, a power cable and a sextant adapter which allows the camera to film through the navigation sextant optical system.

Cameras stowed in the lunar module are two 70mm Hasselblad data cameras fitted with 60mm Zeiss Metric lenses, an electric Hasselblad with 500mm lens and two 16mm Maurer motion picture cameras with 10mm lenses. One of the Hasselblads and one of the motion picture cameras are stowed in the modular equipment stowage assembly (MESA) in the LM descent stage.

The LM Hasselblads have crew chest mounts that fit dove-tail brackets on the crewman's remote control unit, thereby leaving both hands free. One of the LM motion picture cameras will be mounted in the right-hand window to record descent, landing, ascent and rendezvous. The 16mm camera stowed in the MESA will be carried aboard the lunar roving vehicle to record portions of the three EVAs.

Descriptions of the 24-inch panoramic camera and the 3-inch mapping/stellar camera are in the orbital science section of this press kit.

TV and Photographic Equipment Table

Nomenclature	CSM at launch	LM at launch	CM to LM	LM to CM	CM at entry
TV, color, zoom lens (monitor with CM system)	1	1			1
Camera, 35mm Nikon	1				1
Lens - 55mm	1				1
Cassette, 35mm	4				4
Camera, Data Acquisition, 16mm	1	1			1
Lens - 10mm	1	1			1
- 18mm	1				1
- 75mm	1				1
Film magazines	12				12
Camera, lunar surface, 16mm		1			
Battery operated lens - 10mm		1			
magazines	10		10	10	10
Camera, Hasselblad, 70mm Electric	1				1
lens - 80mm	1				1
- 250mm	1				1
- 105mm UV (4 band-pass filters)	1				1
film magazines	6				6
film magazine, 70mm UV	1				1
Camera, Hasselblad, 70mm lunar surface electric		3			
lens - 60mm		2			
- 500mm		1			
film magazines	13		13	13	13
Camera, 24-in. Panoramic (in sim)	1				
film cassette (EVA transfer)	1				1
Camera, 3-in. mapping stellar (sim)	1				
film magazine (EVA transfer)	1				1

ASTRONAUT EQUIPMENT

Space Suit

Apollo crewmen wear two versions of the Apollo space suit: the command module pilot version (CMP-A-7LB) for intravehicular operations in the command module and for extravehicular operations during SIM Bay film retrieval during transearth coast; and the extravehicular version (EV-A-7LB) worn by the commander and lunar module pilot for lunar surface EVAs. The CMP-A-7LB is the EV-A-7L suit used on Apollo 14, except as modified to eliminate lunar surface operations features not needed for Apollo 15 CMP functions and to alter suit fittings to interface with the Apollo 15 spacecraft.

The EV-A-7LB suit differs from earlier Apollo suits by having a waist joint that allows greater mobility while the suit is pressurized--stooping down for setting up lunar surface experiments, gathering samples and for sitting on the lunar roving vehicle.

From the inside out, the integrated thermal meteroid garment worn by the commander and lunar module pilot starts with rubber-coated nylon and progresses outward with layers of nonwoven Dacron, aluminized Mylar film and Beta marquisette for thermal radiation protection and thermal spacers, and finally with a layer of nonflammable Teflon-coated Beta cloth and an abrasion-resistant layer of Teflon fabric--a total of 18 layers.

Both types of the A-7LB suit have a central portion called a torso limb suit assembly consisting of a gas-retaining pressure bladder and an outer structural restraint layer.

The space suit, liquid cooling garment, portable life support system (PLSS), oxygen purge system, lunar extravehicular visor assembly, gloves and lunar boots make up the extravehicular mobility unit (EMU). The EMU provides an extravehicular crewman with life support for a seven-hour mission outside the lunar module without replenishing expendables.

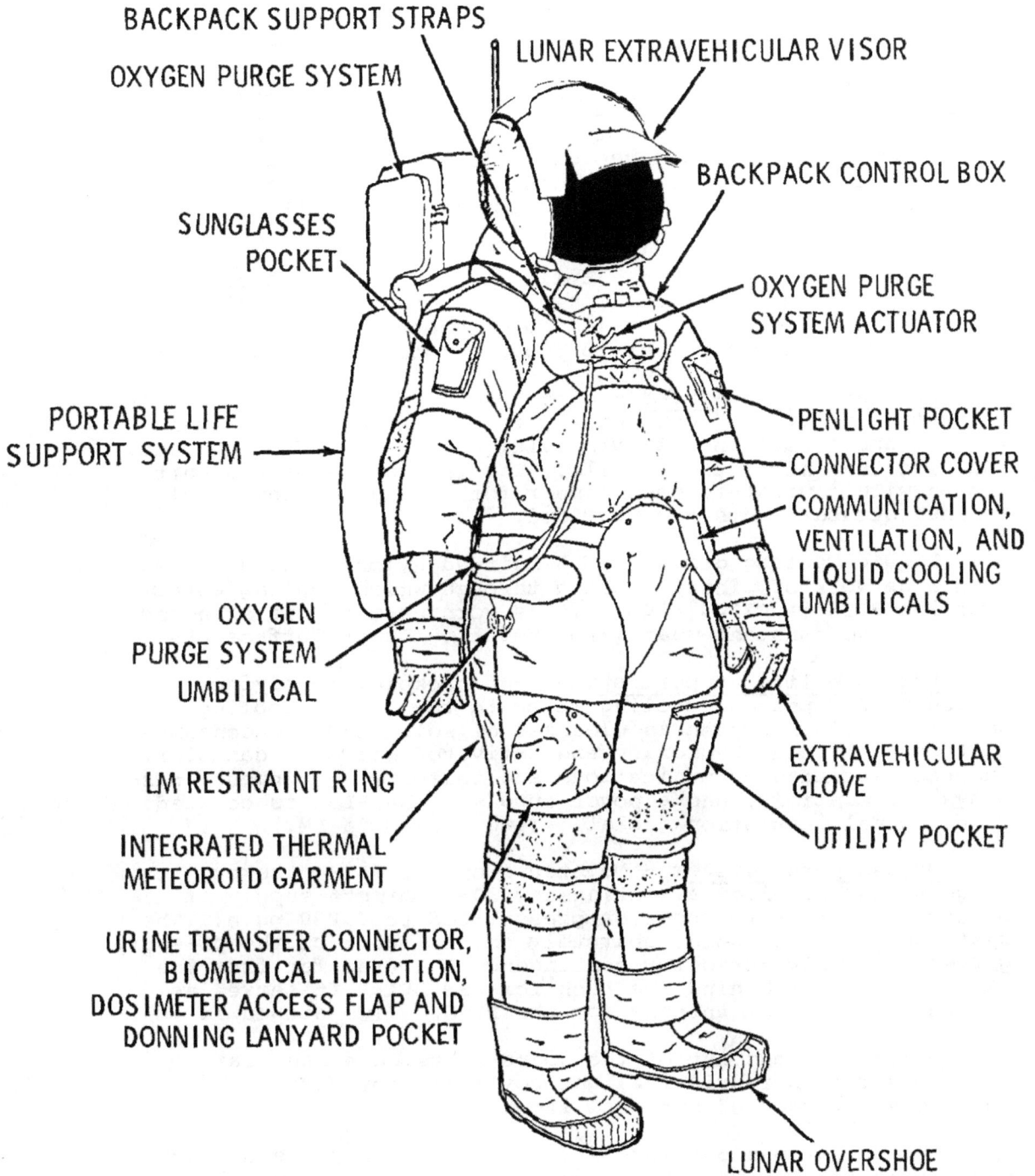

BACKPACK SUPPORT STRAPS

OXYGEN PURGE SYSTEM

LUNAR EXTRAVEHICULAR VISOR

BACKPACK CONTROL BOX

SUNGLASSES POCKET

OXYGEN PURGE SYSTEM ACTUATOR

PORTABLE LIFE SUPPORT SYSTEM

PENLIGHT POCKET

CONNECTOR COVER

COMMUNICATION, VENTILATION, AND LIQUID COOLING UMBILICALS

OXYGEN PURGE SYSTEM UMBILICAL

LM RESTRAINT RING

EXTRAVEHICULAR GLOVE

UTILITY POCKET

INTEGRATED THERMAL METEOROID GARMENT

URINE TRANSFER CONNECTOR, BIOMEDICAL INJECTION, DOSIMETER ACCESS FLAP AND DONNING LANYARD POCKET

LUNAR OVERSHOE

EXTRAVEHICULAR MOBILITY UNIT

Lunar extravehicular visor assembly - A polycarbonate
shell and two visors with thermal control and optical coatings
on them. The EVA visor is attached over the pressure helmet
to provide impact, micrometeoroid, thermal and ultraviolet-
infrared light protection to the EVA crewmen. After Apollo 12,
a sunshade was added to the outer portion of the LEVA in the
middle portion of the helmet rim.

Extravehicular gloves - Built of an outer shell of
Chromel-R fabric and thermal insulation to provide protection
when handling extremely hot and cold objects. The finger tips
are made of silicone rubber to provide more sensitivity.

Constant-wear garment - A one-piece constant-wear
garment, similar to "long johns", is worn as an undergarment
for the space suit in intravehicular and on CSM EV operations,
and with the inflight coveralls. The garment is porous-knit
cotton with a waist-to neck zipper for donning. Biomedical
harness attach points are provided.

Liquid Cooling garment - A knitted nylon-spandex garment
with a network of plastic tubing through which cooling water
from the PLSS is circulated. It is worn next to the skin and
replaces the constant-wear garment during Lunar Surface EVA.

Portable life support system - A backpack supplying
oxygen at 3.7 psi and cooling water to the liquid cooling
garment. Return oxygen is cleansed of solid and gas contaminants by a lithium hydroxide and activated charcoal canister.
The PLSS includes communications and telemetry equipment, displays and controls, and a power supply. The PLSS is covered
by a thermal insulation jacket. (two stowed in LM.)

Oxygen purge system - Mounted atop the PLSS, the oxygen
purge system provides a contingency 30-75 minute supply of
gaseous oxygen in two bottles pressurized to 5,880 psia. The
system may also be worn separately on the front of the pressure
garment assembly torso for contingency EVA transfer from the
LM to the CSM or behind the neck for CSM EVA. It serves as
a mount for the VHF antenna for the PLSS. (Two stowed in LM).

During periods out of the space suits, crewmen wear
two-piece Teflon fabric inflight coveralls for warmth and
for pocket stowage of personal items.

Communications carriers ("Snoopy Hats") with redundant
microphones and earphones are worn with the pressure helmet;
a light-weight headset is worn with the inflight coveralls.

SPACESUIT/PLSS
APOLLO 15 - MAJOR DIFFERENCES

A7L-B SUIT

● ADDED MOBILITY - WAIST CONVOLUTE ADDED
  - NECK CONVOLUTE ADDED
  - ZIPPER RELOCATED FROM CROTCH - AIDS LEG ACTION
  - LESS SHOULDER FORCE

● DURABILITY/PERFORMANCE - FULL BLADDER/CONVOLUTE ABRASION PROTECTION
  (INCREASED EVA CAPABILITY)
  - IMPROVED ZIPPER
  - IMPROVED THERMAL GARMENT
  - INCREASED DRINKING H_2O SUPPLY

-7 PLSS

● LONGER EVA CAPABILITY - INCREASED O_2 SUPPLY
  - INCREASED H_2O SUPPLY
  - LARGER BATTERY
  - ADDITIONAL LiOH

OPS/BSLSS

● NO CHANGE

-more-

APO 1496

PLSS EXPENDABLES COMPARISON

	APOLLO 14	APOLLO 15
OXYGEN	1020 PSIA	1430 PSIA
FEEDWATER	8.50 POUNDS	11.50 POUNDS
BATTERY	279 WATT-HOURS	390 WATT-HOURS
LiOH	3.00 POUNDS	3.12 POUNDS

Quart drinking water bags are attached to the inside neck rings of the EVA suits. The crewman can take a sip of water from the 6-by-8-inch bag through a 1/8-inch-diameter tube within reach of his mouth. The bags are filled from the lunar module potable water dispenser.

Buddy Secondary Life Support System - A connecting hose system which permits a crewman with a failed PLSS to share cooling water in the other crewman's PLSS. Flown for the first time on Apollo 14, the BSLSS lightens the load on the oxygen purge system in the event of a total PLSS failure in that the OPS would supply breathing and pressurizing oxygen while the metabolic heat would be removed by the shared cooling water from the good PLSS. The BSLSS will be stowed on the LRV.

Lunar Boots

The lunar boot is a thermal and abrasion protection device worn over the inner garment and boot assemblies. It is made up of layers of several different materials beginning with teflon coated Beta cloth for the boot liner to Chromel R metal fabric for the outer shell assembly. Aluminized Mylar, Nomex felt, Dacron, Beta cloth and Beta marquisette Kapton comprise the other layers. The lunar boot sole is made of high-strength silicone rubber.

Crew Food System

The Apollo 15 crew selected menus from a list of 100 food items qualified for flight. The balanced menus provide approximately 2,300 calories per man per day. Food packages are assembled into man-meal units for the first ten days of the mission. Items similar to those in the daily menu have been stowed in a pantry fashion which gives the crew some variety in making "real-time" food selection for later meals, snacks and beverages. Also, it allows the crew to supplement or substitute food items contained in the nominal man-meal package.

There are various types of food used in the menus. These include freeze-dried rehydratables in spoon-bowl packages; thermostabilized foods (wet packs) in flexible packages and metal easy-open cans, intermediate moisture and dry bite size cubes and beverages. New food items for this mission are thermostabilized beef steaks and hamburgers, an intermediate moisture apricot food bar and citrus flavored beverage.

Water for drinking and rehydrating food is obtained from two sources in the command module - a portable dispenser for drinking water and a water spigot at the food preparation station which supplies water at about 145 degrees and 55 degrees Fahrenheit. The portable water dispenser provides a continuous flow of water as long as the trigger is held down, and the food preparation spigot dispenses water in one-ounce increments.

A continuous flow water dispenser similar to the one in the command module is used aboard the lunar module for cold-water reconstitution of food stowed aboard the LM.

Water is injected into a food package and the package is kneaded and allowed to sit for several minutes. The bag top is then cut open and the food eaten with a spoon. After a meal, germicide tablets are placed in each bag to prevent fermentation and gas formation. The bags are then rolled and stowed in waste disposal areas in the spacecraft.

Personal Hygiene

Crew personal hygiene equipment aboard Apollo 15 includes body cleanliness items, the waste management system and one medical kit.

Packaged with the food are a toothbrush and a two-ounce tube of toothpaste for each crewman. Each man-meal package contains a 3.5-by-4-inch wet-wipe cleansing towel. Additionally, three packages of 12-by-12-inch dry towels are stowed beneath the command module pilot's couch. Each package contains seven towels. Also stowed under the command module pilot's couch are seven tissue dispensers containing 53 three-ply tissues each.

Solid body wastes are collected in plastic defecation bags which contain a germicide to prevent bacteria and gas formation. The bags are sealed after use and stowed in empty food containers for post-flight analysis.

Urine collection devices are provided for use while wearing either the pressure suit or the inflight coveralls. The urine is dumped overboard through the spacecraft urine dump valve in the CM and stored in the LM.

Medical Kit

The 5-by-5-by-8-inch medical accessory kit is stowed in a compartment on the spacecraft right side wall beside the lunar module pilot couch. The medical kit contains three motion sickness injectors, three pain suppression injectors, one two-ounce bottle first aid ointment, two one-ounce bottles of eye drops, three bottles of nasal drops, two compress bandages, 12 adhesive bandages, one oral thermometer, and four spare crew biomedical harnesses. Pills in the medical kit are 60 antibiotic, 12 nausea, 12 stimulant, 18 pain killer, 60 decongestant, 24 diarrhea, 72 aspirin and 40 antacid. Additionally, a small medical kit containing four stimulant, eight diarrhea and four pain killer pills, 12 aspirin, one bottle eye drops, two compress bandages, eight decongestant pills, one automatic injector containing a pain killer, one bottle nasal drops is stowed in the lunar module flight data file compartment.

Survival Kit

The survival kit is stowed in two CM rucksacks in the right-hand forward equipment bay above the lunar module pilot.

Contents of rucksack No. 1 are: two combination survival lights, one desalter kit, three pairs of sunglasses, one radio beacon, one spare radio beacon battery and spacecraft connector cable, one knife in sheath, three water containers, two containers of Sun lotion, two utility knives, three survival blankets and one utility netting.

Rucksack No. 2: one three-man life raft with CO_2 inflater, one sea anchor, two sea dye markers, three sunbonnets, one mooring lanyard, three manlines and two attach brackets.

The survival kit is designed to provide a 48-hour postlanding (water or land) survival capability for three crewmen between 40 degrees North and South latitudes.

-more-

NATIONAL AERONAUTICS AND SPACE ADMINISTRATION

WASHINGTON, D. C. 20546

BIOGRAPHICAL DATA

NAME: David R. Scott (Colonel, USAF) Apollo 15
    Commander
NASA Astronaut

BIRTHPLACE AND DATE: Born June 6, 1932, in San Antonio, Texas.
    His parents, Brigadier General (USAF Retired) and Mrs.
    Tom W. Scott, reside in La Jolla, California.

PHYSICAL DESCRIPTION: Blond hair; blue eyes; height: 6 feet;
    weight: 175 pounds.

EDUCATION: Graduated from Western High School, Washington, D.C.;
    received a Bachelor of Science from the United States
    Military Academy and the degrees of Master of Science in
    Aeronautics and Astronautics and Engineer in Aeronautics
    and Astronautics from the Massachusetts Institute of
    Technology.

MARITAL STATUS: Married to the former Ann Lurton Ott of San
    Antonio, Texas. Her parents are Brigadier General (USAF
    Retired) and Mrs. Isaac W. Ott of San Antonio.

CHILDREN: Tracy L., March 25, 1961; Douglas W., October 8, 1963.

RECREATIONAL INTERESTS: His hobbies are swimming, handball,
    skiing, and photography.

ORGANIZATIONS: Associate Fellow of the American Institute of
    Aeronautics and Astronautics; and member of the Society
    of Experimental Test Pilots, and Tau Beta Pi, Sigma Xi
    and Sigma Gamma Tau.

SPECIAL HONORS: Awarded the NASA Distinguished Service Medal,
    the NASA Exceptional Service Medal, the Air Force
    Distinguished Service Medal, the Air Force Command Pilot
    Astronaut Wings and the Air Force Distinguished Flying
    Cross; and recipient of the AIAA Astronautics Award (1966)
    and the National Academy of Television Arts and Sciences
    Special Trustees Award (1969).

EXPERIENCE: Scott graduated fifth in a class of 633 at West
          Point and subsequently chose an Air Force career.
          He completed pilot training at Webb Air Force Base,
          Texas, in 1955 and then reported for gunnery training
          at Laughlin Air Force Base, Texas, and Luke Air Force
          Base, Arizona.

          He was assigned to the 32nd Tactical Fighter Squadron
          at Soesterberg Air Base (RNAF), Netherlands, from
          April 1956 to July 1960. Upon completing this tour of
          duty, he returned to the United States for study at the
          Massachusetts Institute of Technology where he completed
          work on his Master's degree. His thesis at MIT concerned
          interplanetary navigation. After completing his studies
          at MIT in June 1962, he attended the Air Force Experimental
          Test Pilot School and then the Aerospace Research Pilot
          School.

          He has logged more than 4,721 hours flying time--4,011 hours
          in jet aircraft and 188 hours in helicopters.

CURRENT ASSIGNMENT: Colonel Scott was one of the third group
          of astronauts named by NASA in October 1963.

          On March 16, 1966, he and command pilot Neil Armstrong
          were launched into space on the Gemini 8 mission--a flight
          originally scheduled to last three days but terminated
          early due to a malfunctioning OAMS thruster. The crew
          performed the first successful docking of two vehicles
          in space and demonstrated great piloting skill in over-
          coming the thruster problem and bringing the spacecraft
          to a safe landing.

          He served as command module pilot for Apollo 9, March 3-13,
          1969. This was the third manned flight in the Apollo
          series and the second to be launched by a Saturn V. The
          ten-day flight encompassed completion of the first com-
          prehensive Earth-orbital qualification and verification
          tests of a "fully configured Apollo spacecraft" and pro-
          vided vital information previously not available on the
          operational performance, stability and reliability of
          lunar module propulsion and life support systems.

          Following a Saturn V launch into a near circular 102.3 x
          103.9 nautical mile orbit, Apollo 9 successfully
          accomplished command/service module separation, trans-
          position and docking maneuvers with the S-IVB-housed lunar
          module. The crew then separated their docked spacecraft
          from the S-IVB third stage and commenced an intensive
          five days of checkout operations with the lunar module,
          followed by five days of command/service module Earth
          orbital operations.

Highlight of this evaluation was completion of a
critical lunar-orbit rendezvous simulation and subsequent docking, initiated by James McDivitt and Russell
Schweickart from within the lunar module at a separation
distance which exceeded 100 miles from the command/service
module piloted by Scott.

The crew also demonstrated and confirmed the operational
feasibility of crew transfer and extravehicular activity
techniques and equipment, with Schweickart completing
a 46-minute EVA outside the lunar module. During this
period, Dave Scott completed a stand-up EVA in the open
command module hatch photographing Schweickart's activities
and also retrieving thermal samples from the command
module exterior.

Apollo 9 splashed down less than four miles from the
helicopter carrier USS GUADALCANAL. With the completion
of this flight, Scott has logged 251 hours and 42 minutes
in space.

He served as backup spacecraft commander for the Apollo
12 flight and is currently assigned as spacecraft commander
for Apollo 15.

-end-

June 1971

NATIONAL AERONAUTICS AND SPACE ADMINISTRATION

WASHINGTON, D. C. 20546

BIOGRAPHICAL DATA

NAME: Alfred Merrill Worden (Major, USAF) Apollo 15
     Command Module Pilot
    NASA Astronaut

BIRTHPLACE AND DATE: The son of Merrill and Helen Worden, he
    was born in Jackson, Michigan, on February 7, 1932.
    His parents reside in Jackson, Michigan.

PHYSICAL DESCRIPTION: Brown hair; blue eyes; height: 5 feet
    10 1/2 inches; weight: 153 pounds.

EDUCATION: Attended Dibble, Griswold, Bloomfield and East
    Jackson grade schools and completed his secondary
    education at Jackson High School; received a Bachelor
    of Military Science Degree from the United States Military
    Academy in 1955 and Master of Science degrees in
    Astronautical/Aeronautical Engineering and Instrumentation
    Engineering from the University of Michigan in 1963.

CHILDREN: Merrill E., January 16, 1958; Alison P., April 6, 1960.

RECREATIONAL INTERESTS: He enjoys bowling, water skiing,
    swimming and handball.

EXPERIENCE: Worden, an Air Force Major, was graduated from the
    United States Military Academy in June 1955 and, after
    being commissioned in the Air Force, received flight
    training at Moore Air Base, Texas; Laredo Air Force Base,
    Texas; and Tyndall Air Force Base, Florida.

    Prior to his arrival for duty at the Manned Spacecraft
    Center, he served as an instructor at the Aerospace
    Research Pilots School--from which he graduated in
    September 1965. He is also a graduate of the Empire Test
    Pilots School in Farnborough, England, and completed his
    training there in February 1965.

    He attended Randolph Air Force Base Instrument Pilots
    Instructor School in 1963 and served as a pilot and
    armament officer from March 1957 to May 1961 with the 95th
    Fighter Interceptor Squadron at Andrews Air Force Base,
    Maryland.

-more-

He has logged more than 3,309 hours flying time--
including 2,804 hours in jets and 107 in helicopters.

CURRENT ASSIGNMENT: Major Worden is one of the 19 astronauts
selected by NASA in April 1966. He served as a member
of the astronaut support crew for the Apollo 9 flight
and as backup command module pilot for the Apollo 12
flight.

He is currently assigned as command module pilot for
Apollo 15.

-end-

June 1971

NATIONAL AERONAUTICS AND SPACE ADMINISTRATION

WASHINGTON, D. C. 20546

BIOGRAPHICAL DATA

NAME: James Benson Irwin (Lieutenant Colonel, USAF) Apollo 15
Lunar Module Pilot
NASA Astronaut

BIRTHPLACE AND DATE: Born March 17, 1930, in Pittsburgh,
Pennsylvania, but he considers Colorado Springs, Colorado,
as his home town. His parents, Mr. and Mrs. James Irwin,
now reside in San Jose, California.

PHYSICAL DESCRIPTION: Brown hair; brown eyes; height: 5 feet
8 inches; weight: 160 pounds.

EDUCATION: Graduated from East High School, Salt Lake City, Utah;
received a Bachelor of Science degree in Naval Sciences
from the United States Naval Academy in 1951 and Master
of Science degrees in Aeronautical Engineering and
Instrumentation Engineering from the University of
Michigan in 1957.

MARITAL STATUS: Married to the former Mary Ellen Monroe of
Corvallis, Oregon; her parents, Mr. and Mrs. Leland F.
Monroe, reside in Santa Clara, California.

CHILDREN: Joy C., November 26, 1959; Jill C., February 22, 1961;
James B., January 4, 1963; Jan C., September 30 1964.

RECREATIONAL INTERESTS: Enjoys skiing and playing paddleball,
handball, and squash; and his hobbies include fishing,
diving, and camping.

ORGANIZATIONS: Member of the Air Force Association and the
Society of Experimental Test Pilots.

SPECIAL HONORS: Winner of two Air Force Commendation Medals for
service with the Air Force Systems Command and the Air
Defense Command; and, as a member of the 4750th Training
Wing, recipient of an Outstanding Unit Citation.

-more-

EXPERIENCE: Irwin, an Air Force Lt. Colonel, was commissioned
in the Air Force on graduation from the Naval Academy in
1951. He received his flight training at Hondo Air Base,
Texas, and Reese Air Force Base, Texas.

Prior to reporting for duty at the Manned Spacecraft Center,
he was assigned as Chief of the Advanced Requirements Branch
at Headquarters Air Defense Command. He was graduated from
the Air Force Aerospace Research Pilot School in 1963 and
the Air Force Experimental Test Pilot School in 1961.

He also served with the F-12 Test Force at Edwards Air
Force Base, California, and with the AIM 47 Project
Office at Wright-Patterson Air Force Base, Ohio.

During his military career, he has accumulated more than
6,650 hours flying time--5,124 hours in jet aircraft and
387 in helicopters.

CURRENT ASSIGNMENT: Lt. Colonel Irwin is one of the 19 astronauts
selected by NASA in April 1966. He was crew commander of
lunar module (LTA-8)--this vehicle finished the first
series of thermal vacuum tests on June 1, 1968. He also
served as a member of the astronaut support crew for
Apollo 10 and as backup lunar module pilot for the Apollo 12
flight.

Irwin is currently assigned as lunar module pilot for
Apollo 15.

-end-

June 1971

NATIONAL AERONAUTICS AND SPACE ADMINISTRATION

WASHINGTON, D. C. 20546

BIOGRAPHICAL DATA

NAME: Richard F. Gordon, Jr. (Captain, USN), Backup
       Apollo 15 Commander
       NASA Astronaut

BIRTHPLACE AND DATE: Born October 5, 1929, in Seattle,
Washington. His mother, Mrs. Angela Gordon, resides in
Seattle.

PHYSICAL DESCRIPTION: Brown hair; hazel eyes; height: 5
feet 7 inches; weight: 150 pounds.

EDUCATION: Graduated from North Kitsap High School,
Poulsbo, Washington; received a Bachelor of Science degree
in Chemistry from the University of Washington in 1951.

MARITAL STATUS: Married to the former Barbara J. Field
of Seattle, Washington. Her parents, Mr. and Mrs. Chester
Field, reside in Freeland, Washington.

CHILDREN: Carleen, July 8, 1954; Richard, October 6, 1955;
Lawrence, December 18, 1957; Thomas, March 25, 1959; James,
April 26, 1960; Diane, April 23, 1961.

RECREATIONAL INTERESTS: He enjoys water skiing, sailing,
and golf.

ORGANIZATIONS: Member of the Society of Experimental Test
Pilots.

SPECIAL HONORS: Awarded two Navy Distinguished Flying
Crosses, the NASA Exceptional Service Medal, the Navy Astronaut Wings, the Navy Distinguished Service Medal, the Institute of Navigation Award for 1969, the Godfrey L. Cabot
Award in 1970, and the Rear Admiral William S. Parsons Award
for Scientific and Technical Progress in 1970.

EXPERIENCE: Gordon, a Navy Captain, received his wings as
a naval aviator in 1953. He then attended All-weather Flight
School and jet transitional training and was subsequently
assigned to an all-weather fighter squadron at the Naval Air
Station at Jacksonville, Florida.

-more-

In 1957, he attended the Navy's Test Pilot School at Patuxent River, Maryland, and served as a flight test pilot until 1960. During this tour of duty, he did flight test work on the F8U Crusader, F11F Tigercat, FJ Fury, and A4D Skyhawk, and was the first project test pilot for the F4H Phantom II.

He served with Fighter Squadron 121 at the Miramar, California, Naval Air Station as a flight instructor in the F4H and participated in the introduction of that aircraft to the Atlantic and Pacific fleets. He was also flight safety officer, assistant operations officer, and ground training officer for Fighter Squadron 96 at Miramar.

Winner of the Bendix Trophy Race from Los Angeles to New York in May 1961, he established a new speed record of 869.74 miles per hour and a transcontinental speed record of two hours and 47 minutes.

He was also a student at the U.S. Naval Postgraduate School at Monterey, California.

He has logged more than 4,682 hours flying time--3,775 hours in jet aircraft and 121 in helicopters.

CURRENT ASSIGNMENT: Captain Gordon was one of the third group of astronauts named by NASA in October 1963. He served as backup pilot for the Gemini 8 flight.

On September 12, 1966, he served as pilot for the 3-day Gemini 11 mission--on which rendezvous with an Agena was achieved in less than one orbit. He executed docking maneuvers with the previously launched Agena and performed two periods of extravehicular activity which included attaching a tether to the Agena and retrieving a nuclear emulsion experiment package. Other highlights accomplished by Gordon and command pilot Charles Conrad on this flight included the successful completion of the first tethered station-keeping exercise, establishment of a new altitude record of 850 miles, and completion of the first fully automatic controlled reentry. The flight was concluded on September 15, 1966, with the spacecraft landing in the Atlantic--2 1/2 miles from the prime recovery ship USS GUAM.

Gordon was subsequently assigned as backup command module pilot for Apollo 9.

He occupied the command module pilot seat on Apollo 12,
November 14-24, 1969. Other crewmen on man's second lunar
landing mission were Charles Conrad (spacecraft commander)
and Alan L. Bean (lunar module pilot). Throughout the 31-
hour lunar surface stay by Conrad and Bean, Gordon remained
in lunar orbit aboard the command module, "Yankee Clipper,"
obtaining desired mapping photographs of tentative landing
sites for future missions. He also performed the final re-
docking maneuvers following the successful lunar orbit rendezvous which was initiated by Conrad and Bean from within
"Intrepid" after their ascent from the Moon's surface.

All of the mission's objectives were accomplished, and
Apollo 12 achievements include: the first precision lunar
landing with "Intrepid's" touchdown in the Moon's Ocean of
Storms; the first lunar traverse by Conrad and Bean as they
deployed the Apollo Lunar Surface Experiment Package (ALSEP),
installed a nuclear power generator station to provide the
power source for these long-term scientific experiments,
gathered samples of the lunar surface for return to Earth, and
completed a close up inspection of the Surveyor III spacecraft.

The Apollo 12 mission lasted 244 hours and 36 minutes and
was concluded with a Pacific splashdown and subsequent recovery by the USS HORNET.

Captain Gordon has completed two space flights, logging a
total of 315 hours and 53 minutes in space--2 hours and
44 minutes of which were spent in EVA.

He is currently assigned as backup spacecraft commander for
Apollo 15.

-end-

June 1971

NATIONAL AERONAUTICS AND SPACE ADMINISTRATION

WASHINGTON, D. C. 20546

BIOGRAPHICAL DATA

NAME: Vance DeVoe Brand (Mr.), Backup Apollo 15 Command
       Module Pilot
       NASA Astronaut

BIRTHPLACE AND DATE: Born in Longmont, Colorado, May 9,
1931. His parents, Dr. and Mrs. Rudolph W. Brand, reside
in Longmont.

PHYSICAL DESCRIPTION: Blond hair; gray eyes; height: 5 feet
11 inches; weight: 175 pounds.

EDUCATION: Graduated from Longmont High School, Longmont,
Colorado; received a Bachelor of Science degree in Business
from the University of Colorado in 1953, a Bachelor of Science
degree in Aeronautical Engineering from the University of
Colorado in 1960, and a Master's degree in Business Administration from the University of California at Los Angeles in
1964.

MARITAL STATUS: Married to the former Joan Virginia Weninger
of Chicago, Illinois. Her parents, Mr. and Mrs. Ralph D.
Weninger, reside in Chicago.

CHILDREN: Susan N., April 30, 1954; Stephanie, August 6,
1955; Patrick R., March 22, 1958; Kevin S., December 1, 1963.

RECREATIONAL INTERESTS: Skin diving, skiing, handball, and
jogging.

ORGANIZATIONS: Member of the Society of Experimental Test
Pilots, the American Institute of Aeronautics and Astronautics,
Sigma Nu, and Beta Gamma Sigma.

EXPERIENCE: Brand served as a commissioned officer and naval
aviator with the U.S. Marine Corps from 1953 to 1957. His
Marine Corps assignments included a 15-month tour in Japan as
a jet fighter pilot. Following his release from active duty,
he continued flying fighter aircraft in the Marine Corps Reserve
and the Air National Guard until 1964, and he still retains
a commission in the Air Force Reserve.

-more-

From 1960 to 1966, Brand was employed as a civilian by the Lockheed Aircraft Corporation. He first worked as a flight test engineer on the P3A "Orion" aircraft and later transferred to the experimental test pilot ranks. In 1963, he graduated from the U.S. Naval Test Pilot School and was assigned to Palmdale, California, as an experimental test pilot on Canadian and German F-104 development programs. Immediately prior to his selection to the astronaut program, Brand was assigned to the West German F-104G Flight Test Center at Istres, France, as an experimental test pilot and leader of a Lockheed flight test advisory group.

He has logged 3,984 hours of flying time, which include 3,216 in jets and 326 hours in helicopters.

CURRENT ASSIGNMENT: Mr. Brand is one of the 19 astronauts selected by NASA in April 1966. He served as a crew member for the thermal vacuum test of 2TV-1, the prototype command module; and he was a member of the astronaut support crews for the Apollo 8 and 13 missions.

Currently he is backup command module pilot for Apollo 15.

-end-

June 1971

NATIONAL AERONAUTICS AND SPACE ADMINISTRATION

WASHINGTON, D. C. 20546

BIOGRAPHICAL DATA

NAME: Harrison H. Schmitt (PhD), Backup Apollo 15 Lunar
       Module Pilot
       NASA Astronaut

BIRTHPLACE AND DATE: Born July 3, 1935, in Santa Rita,
New Mexico. His mother, Mrs. Harrison A. Schmitt, resides
in Silver City, New Mexico.

PHYSICAL DESCRIPTION: Black hair; brown eyes; height: 5 feet
9 inches; weight: 165 pounds.

EDUCATION: Graduated from Western High School, Silver City,
New Mexico; received a Bachelor of Science degree in Science
from the California Institute of Technology in 1957; studied
at the University of Oslo in Norway during 1957-58; received
Doctorate in Geology from Harvard University in 1964.

MARITAL STATUS: Single.

RECREATIONAL INTERESTS: His hobbies include skiing, hunting,
fishing, carpentry, and hiking.

ORGANIZATIONS: Member of the Geological Society of America,
American Geophysical Union, and Sigma Xi.

SPECIAL HONORS: Winner of a Fulbright Fellowship (1957-58);
a Kennecott Fellowship in Geology (1958-59); a Harvard Fellow-
ship (1959-60); a Harvard Traveling Fellowship (1960); a Parker
Traveling Fellowship (1961-62); a National Science Foundation
Post-Doctoral Fellowship, Department of Geological Sciences,
Harvard University (1963-64).

EXPERIENCE: Schmitt was a teaching fellow at Harvard in
1961; he assisted in the teaching of a course in ore deposits
there. Prior to his teaching assignment, he did geological
work for the Norwegian Geological Survey in Oslo, Norway, and
for the U.S. Geological Survey in New Mexico and Montana. He
also worked as a geologist for two summers in Southeastern
Alaska.

-more-

Before coming to the Manned Spacecraft Center, he served with the U.S. Geological Survey's Astrogeology Branch at Flagstaff, Arizona. He was project chief for lunar field geological methods and participated in photo and telescopic mapping of the Moon; he was among the USGS astrogeologists instructing NASA astronauts during their geological field trips.

He has logged more than 1,329 hours flying time--1,141 hours in jet aircraft and 177 in helicopters.

CURRENT ASSIGNMENT: Dr. Schmitt was selected as a scientist-astronaut by NASA in June 1965. He completed a 53-week course in flight training at Williams Air Force Base, Arizona and, in addition to training for future manned space flights, has been instrumental in providing Apollo flight crews with detailed instruction in lunar navigation, geology, and feature recognition.

Schmitt is currently assigned as backup lunar module pilot for Apollo 15.

-end-

June 1971

APOLLO 15 FLAGS, LUNAR MODULE PLAQUE

The United States flag to be erected on the lunar surface measures 30 by 48 inches and will be deployed on a two-piece aluminum tube eight feet long. The flag, made of nylon, will be stowed in the lunar module descent stage modularized equipment stowage assembly.

Also carried on the mission and returned to Earth will be 25 United States flags, 50 individual state flags, flags of United States territories and flags of all United Nations member nations, each four by six inches.

A seven by nine-inch stainless steel plaque, similar to that flown on Apollo 14 will be fixed to the LM front leg. The plaque has on it the words "Apollo 15" with "Falcon" beneath, "July 1971," and the signatures of the three crewmen.

SATURN V LAUNCH VEHICLE

The Saturn V launch vehicle (SA-510) assigned to the Apollo 15 mission was developed under the direction of the Marshall Space Flight Center, Huntsville, Ala. The vehicle is similar to those vehicles used for the missions of Apollo 8 through Apollo 14.

First Stage

The first stage (S-1C) of the Saturn V was build by the Boeing Co. at NASA's Michoud Assembly Facility, New Orleans. The stage's five F-1 engines develop about 7.7 million pounds of thrust at launch. Major components of the stage are the forward skirt, oxidizer tank, inter-tank structure, fuel tank, and thrust structure. Propellant to the five engines normally flows at a rate of approximately 29,400 pounds (3,400 gallons) a second. One engine is rigidly mounted on the stage's centerline; the outer four engines are mounted on a ring at 90-degree angles around the center engine. These outer engines are gimbaled to control the vehicle's attitude during flight.

Second Stage

The second stage (S-II) was built by the Space Division of the North American Rockwell Corp. at Seal Beach, Calif. Five J-2 engines develop a total of about 1.15 million pounds of thrust during flight. Major structural components are the forward skirt, liquid hydrogen and liquid oxygen tanks (separated by an insulated common bulkhead), a thrust structure and an interstage section that connects the first and second stages. The engines are mounted and used in the same arrangement as the first stage's F-1 engines: four outer engines can be gimbaled; the center one is fixed.

Third Stage

The third stage (S-IVB) was built by the McDonnell Douglas Astronautics Co. at Huntington Beach, Calif. Major components are the aft interstage and skirt, thrust structure, two propellant tanks with a common bulkhead, a forward skirt, and a single J-2 engine. The gimbaled engine has a maximum thrust of 230,000 pounds, and can be restarted in Earth orbit.

-more-

SPACECRAFT 82 FT.

CM

SM

LM INSTRUMENT UNIT

THIRD STAGE (S-IVB)

SECOND STAGE (S-II)

FIRST STAGE (S-IC)

SATURN V LAUNCH VEHICLE -281 FT.

INSTRUMENT UNIT (IU)

Diameter:	21.7 feet
Height:	3 feet
Weight:	4,500 lbs.

THIRD STAGE (S-IVB)

Diameter:	21.7 feet
Height:	59.3 feet
Weight:	260,000 lbs. fueled
	25,000 lbs. dry
Engine:	One J-2
Propellants:	Liquid Oxygen (189,800 lbs.; 20,000 gals.)
	Liquid Hydrogen (43,500 lbs.; 74,150 gals.)
Thrust:	198,800 lbs. to 190,500 lbs.
Interstage:	8,000 lbs.

SECOND STAGE (S-II)

Diameter:	33 feet
Height:	81.5 feet
Weight:	1,101,000 lbs. fueled
	78,000 lbs. dry
Engines:	Five J-2
Propellants:	Liquid Oxygen (837,200 lbs.; 88,200 gals.)
	Liquid Hydrogen (159,700 lbs.; 272,200 gals.)
Thrust:	1,150,000 lbs.
Interstage:	11,400 lbs.

FIRST STAGE (S-IC)

Diameter:	33 feet
Height:	138 feet lbs.
Weight:	4,930,000 fueled
	289,800 lbs. dry
Engines:	Five F-1
Propellants:	Liquid Oxygen (3,306,000 lbs.; 348,300 gals.)
	RP-1 Kerosene (1,438,000 lbs.; 215,700 gals.)
Thrust:	7,766,000 lbs. at liftoff

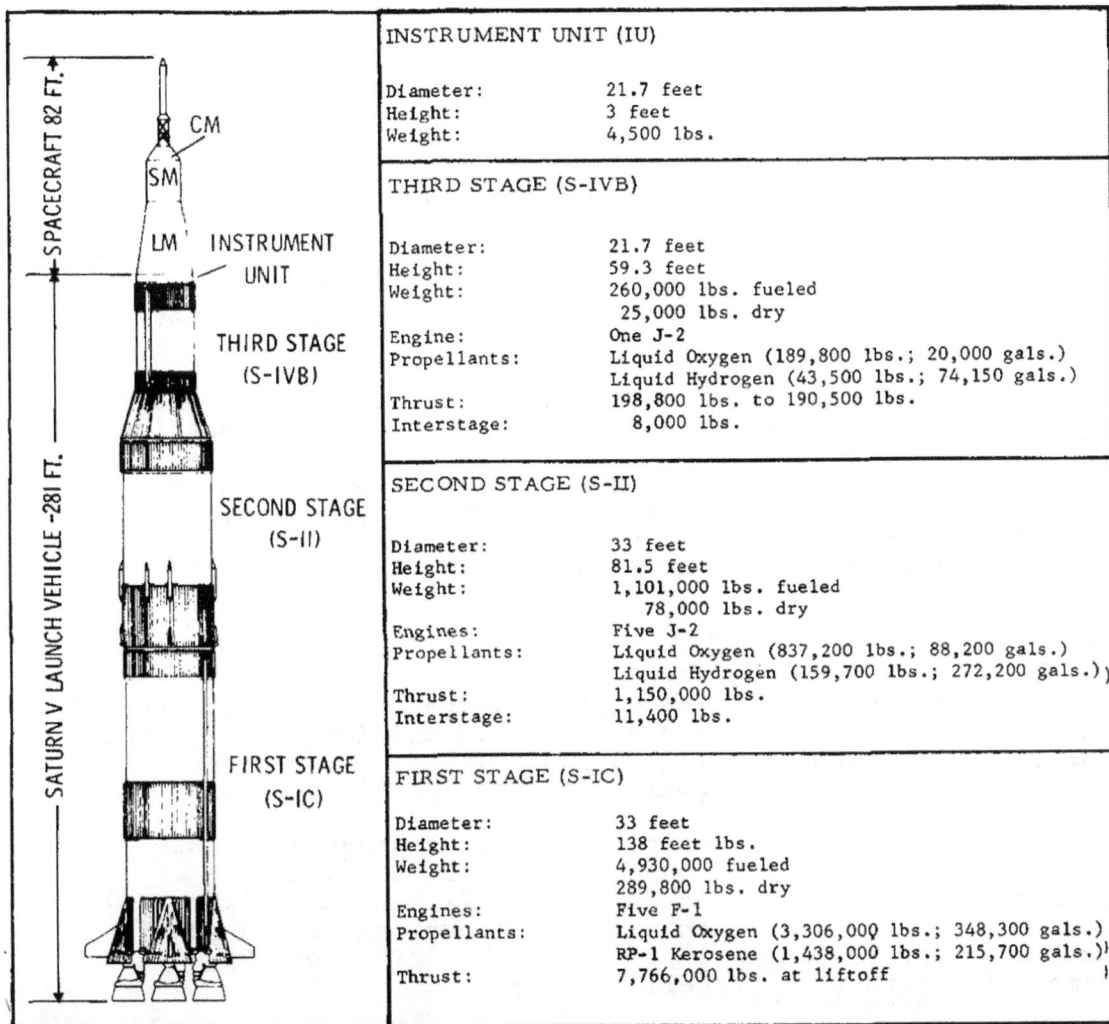

NOTE: Weights and measures given above are for the nominal vehicle configuration for Apollo. The figures may vary slightly due to changes before launch to meet changing conditions. Weights of dry stages and propellants do not equal total weight because frost and miscellaneous smaller items are not included in chart.

SATURN V LAUNCH VEHICLE

-more-

Instrument Unit

The instrument unit (IU), built by the International Business Machines Corp. at Huntsville, Ala., contains navigation, guidance, and control equipment to steer the launch vehicle into Earth orbit and into translunar trajectory. The six major systems are structural, environmental control, guidance and control, measuring and telemetry, communications, and electrical.

The instrument unit's inertial guidance platform provides space-fixed reference coordinates and measures acceleration along three mutually perpendicular axes of a coordinate system. In the unlikely event of platform failure during boost, systems in the Apollo spacecraft are programmed to provide guidance for the launch vehicle. After second stage ignition, the spacecraft commander can manually steer the vehicle if the launch vehicle's stable platform were lost.

Propulsion

The Saturn V has 27 propulsive units, with thrust ratings ranging from 70 pounds to more than 1.5 million pounds. The large main engines burn liquid propellants; the smaller units use solid or hypergolic (self-igniting) propellants.

The five F-1 engines give the first stage a thrust range of from 7,765,852 pounds at liftoff to 9,155,147 pounds at center engine cutoff. Each F-1 engine weighs almost 10 tons, is more than 18 feet long, and has a nozzle exit diameter of nearly 14 feet. Each engine consumes almost three tons of propellant a second.

The first stage has four solid-fuel retro-rockets that fire to separate the first and second stages. Each retro-rocket produces a thrust of 75,800 pounds for 0.54 seconds.

Thrust of the J-2 engines on the second and third stages is 205,000 pounds during flight, operating through a range of 180,000 to 230,000 pounds. The 3,500-pound J-2 engine provides higher thrust for each pound of propellant burned per second than the F-1 engine because the J-2 burns high-energy, low molecular weight, liquid hydrogen. F-1 and J-2 engines are built by the Rocketdyne Division of the North American Rockwell Corp. at Canoga Park, Calif.

Four retro-rockets, located in the S-IVB's aft interstage, separate the S-II from the S-IVB. Two jettisonable ullage rockets settle propellants before engine ignition. Six smaller engines in the two auxiliary propulsion system modules on the S-IVB stage provide three-axis attitude control.

Significant Vehicle Changes

Saturn V vehicle SA-510 will be able to deliver a payload that is more than 4,000 pounds heavier than the Apollo 14 payload. The increase provides for the first Lunar Roving Vehicle and for an exploration time on the lunar surface almost twice that of any other Apollo mission. The payload increases were achieved by revising some operational aspects of the Saturn V and through minor changes to vehicle hardware.

The major operational changes are an Earth parking orbit altitude of 90 nautical miles (rather than 100), and a launch azimuth range of 80 to 100 degrees (rather than 72 to 96). Other operational changes include slightly reduced propellant reserves and increased propellant loading for the first opportunity translunar injection (TLI). A significant portion of the payload increase is due to more favorable temperature and wind effects for a July launch versus one in January.

Most of the hardware changes have been made to the first (S-IC) stage. They include reducing the number of retro-rocket motors (from eight to four), reorificing the F-1 engines, burning the outboard engines nearer to LOX depletion, and burning the center engine longer than before. Another change has been made in the propellant pressurization system of the second (S-II) stage.

Three other changes to the launch vehicle were first made to the Apollo 14 vehicle: a helium gas accumulator is installed in the S-II's center engine liquid oxygen (LOX) line, a backup cutoff device is in the same engine, and a simplified propellant utilization valve is installed on all J-2 engines. These changes prevent high oscillations (the "pogo" effect) in the S-II stage and provide more efficient J-2 engine performance. For Apollo 15 a defective cutoff device can be remotely deactivated on the pad or in flight to prevent an erroneous "vote" for cutoff.

APOLLO SPACECRAFT

The Apollo spacecraft for the Apollo 15 mission consists
of the command module, service module, lunar module, a spacecraft lunar module adapter (SLA), and a launch escape system.
The SLA houses the lunar module and serves as a mating
structure between the Saturn V instrument unit and the SM.

Launch Escape System (LES) -- The function of the LES
is to propel the command module to safety in an aborted
launch. It has three solid-propellant rocket motors: a
147,000-pound-thrust launch escape system motor, a 2,400-
pound-thrust pitch control motor, and a 31,500-pound-thrust
tower jettison motor. Two canard vanes deploy to turn the
command module aerodynamically to an attitude with the heat-
shield forward. The system is 33 feet tall and four feet
in diameter at the base, and weighs 9,108 pounds.

Command Module (CM) -- The command module is a pressure
vessel encased in heat shields, cone-shaped, weighing 12,831
pounds at launch.

The command module consists of a forward compartment
which contains two reaction control engines and components of
the Earth landing system; the crew compartment or inner pressure
vessel containing crew accommodations, controls and displays,
and many of the spacecraft systems; and the aft compartment
housing ten reaction control engines, propellant tankage,
helium tanks, water tanks, and the CSM umbilical cable. The
crew compartment contains 210 cubic feet of habitable volume.

Heat-shields around the three compartments are made of
brazed stainless steel honeycomb with an outer layer of
phenolic epoxy resin as an ablative material.

The CSM and LM are equipped with the probe-and-drogue
docking hardware. The probe assembly is a powered folding
coupling and impact attentuating device mounted in the CM
tunnel that mates with a conical drogue mounted in the LM
docking tunnel. After the 12 automatic docking latches are
checked following a docking maneuver, both the probe and
drogue are removed to allow crew transfer between the CSM
and LM.

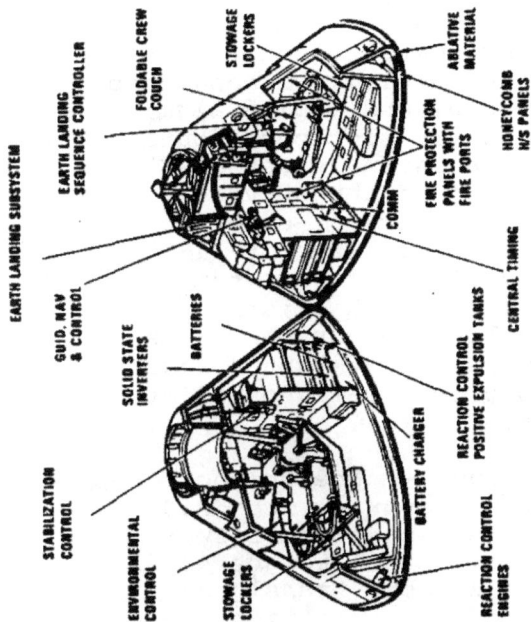

COMMAND MODULE

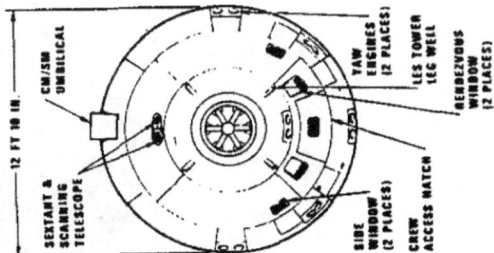

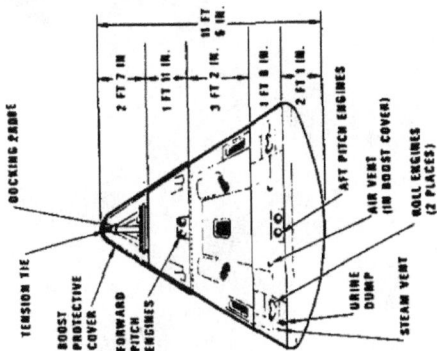

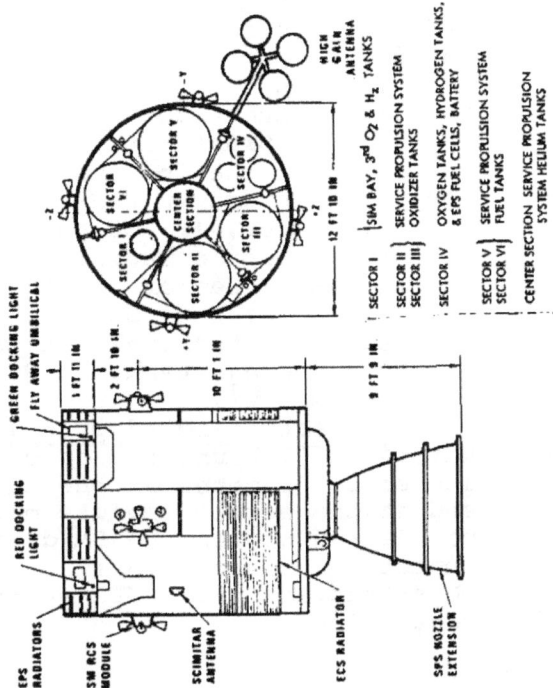

EARTH LANDING SUBSYSTEM

EARTH LANDING SEQUENCE CONTROLLER

FOLDABLE CREW COUCH

STOWAGE LOCKERS

ABLATIVE MATERIAL

FIRE PROTECTION PANELS WITH FIRE PORTS

HONEYCOMB M/S PANELS

GUID. NAV & CONTROL

COMM

CENTRAL TIMING

BATTERIES

SOLID STATE INVERTERS

REACTION CONTROL POSITIVE EXPULSION TANKS

STABILIZATION CONTROL

ENVIRONMENTAL CONTROL

STOWAGE LOCKERS

BATTERY CHARGER

REACTION CONTROL ENGINES

12 FT 10 IN

CM/SM UMBILICAL

SEXTANT & SCANNING TELESCOPE

YAW ENGINES (2 PLACES)

LES TOWER LEG WELL

RENDEZVOUS WINDOW (2 PLACES)

SIDE WINDOW (2 PLACES)

CREW ACCESS HATCH

TENSION TIE

DOCKING PROBE

2 FT 7 IN.

1 FT 11 IN.

11 FT 6 IN.

3 FT 2 IN.

1 FT 8 IN.

2 FT 1 IN.

AFT PITCH ENGINES

AIR VENT (IN BOOST COVER)

ROLL ENGINES (2 PLACES)

BOOST PROTECTIVE COVER

FORWARD PITCH ENGINES

BRINE DUMP

STEAM VENT

SERVICE MODULE

HIGH GAIN ANTENNA

SECTOR I — SM BAY, 3rd O_2 & H_2 TANKS

SECTOR II / SECTOR III — SERVICE PROPULSION SYSTEM OXIDIZER TANKS

SECTOR IV — OXYGEN TANKS, HYDROGEN TANKS, & EPS FUEL CELLS, BATTERY

SECTOR V / SECTOR VI — SERVICE PROPULSION SYSTEM FUEL TANKS

CENTER SECTION — SERVICE PROPULSION SYSTEM HELIUM TANKS

12 FT 10 IN

SECTOR V

SECTOR VI

SECTOR I

CENTER SECTION

SECTOR IV

SECTOR II

SECTOR III

GREEN DOCKING LIGHT

FLY AWAY UMBILICAL

RED DOCKING LIGHT

EPS RADIATORS

SM RCS MODULE

SCIMITAR ANTENNA

ECS RADIATOR

SPS NOZZLE EXTENSION

1 FT 11 IN

2 FT 10 IN

10 FT

9 FT 3 IN

Service Module (SM) -- The Apollo 15 service module will weigh 54,063 pounds at launch, of which 40,593 pounds is propellant for the 20,500-pound thrust service propulsion engine: (fuel: 50/50 hydrazine and unsymmetrical dimethyl-hydrazine; oxidizer: nitrogen tetroxide). Aluminum honeycomb panels one-inch thick form the outer skin, and milled aluminum radial beams separate the interior into six sections around a central cylinder containing service propulsion system (SPS) helium pressurant tanks. The six sectors of the service module house the following components: Sector I-- oxygen tank 3 and hydrogen tank 3, J-mission SIM bay; Sector II--space radiator, +Y RCS package, SPS oxidizer storage tank; Sector III--space radiator, +Z RCS package, SPS oxidizer storage tank; Sector IV--three fuel cells, two oxygen tanks, two hydrogen tanks, auxiliary battery; Sector V--space radiator, SPS fuel sump tank, -Y RCS package; Sector VI--space radiator, SPS fuel storage tank, -Z RCS package.

Spacecraft-LM adapter (SLA) Structure -- The spacecraft-LM adapter is a truncated cone 28 feet long tapering from 260 inches in diameter at the base to 154 inches at the forward end at the service module mating line. The SLA weighs 4,061 pounds and houses the LM during launch and the translunar injection maneuver until CSM separation, transposition, and LM extraction. The SLA quarter panels are jettisoned at CSM separation.

Command-Service Module Modifications

Following the Apollo 13 abort in April 1970, several
changes were made to enhance the capability of the CSM to
return a flight crew safely to Earth should a similar incident
occur again.

These changes included the addition of an auxiliary
storage battery in the SM, the removal of destratification
fans in the cryogenic oxygen tanks, and the removal of
thermostat switches from the oxygen tank heater circuits.
The auxiliary battery added was a 415-ampere hour, silver
oxide/zinc, non-rechargeable type, weighing 135 pounds which
is identical to the five lunar module descent batteries.

Additional changes incorporated were a third 320-pound
capacity cryogenic oxygen tank in the SM (Sector I), a
valve which allows the third oxygen tank to be isolated from
the fuel cells and from the other two tanks in an emergency
so as to feed only the command module environmental control
system. These latter changes had already been planned for
the Apollo 15 J-series spacecraft to extend the mission
duration.

Additionally, a third 26-pound capacity hydrogen tank
was added to complement the third oxygen tank to provide
for additional power.

Other changes to the CSM include addition of handrails
and foot restraints for the command module pilot's transearth coast EVA to retrieve film casettes from the SIM bay
cameras. The SIM bay and its experiment packages, described
in more detail in the orbital science section of the press
kit, have been thermally isolated from the rest of the
service module by the addition of insulation material.

Additional detailed information on command module and
lunar module systems and subsystems is available in reference
documents at query desks at KSC and MSC News Centers.

MODIFIED COMMAND AND SERVICE MODULE
CSM-112 AND SUBSEQUENT

COMMAND MODULE
CONTROLS FOR SIM EXPERIMENTS
EVA CAPABILITY

SERVICE MODULE
H2 TANK
EXPANDED DATA SYSTEM
JETTISONABLE PANEL
SIM AND INSTRUMENTS
EVA CAPABILITY

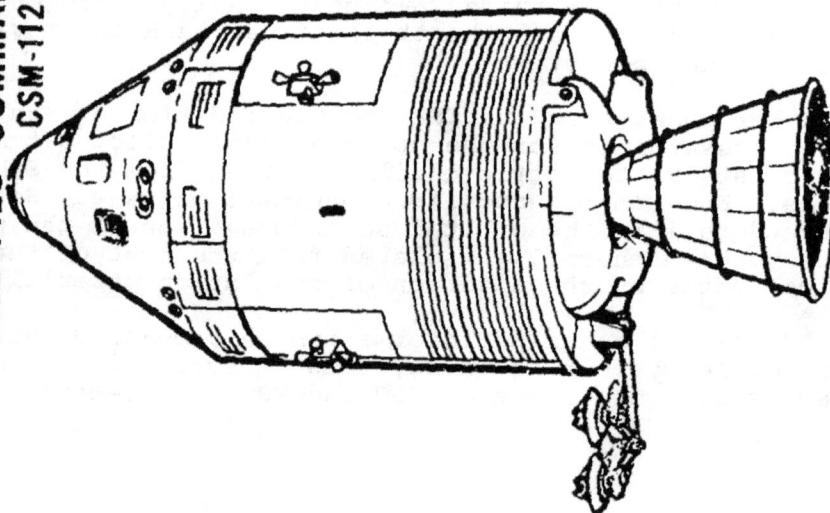

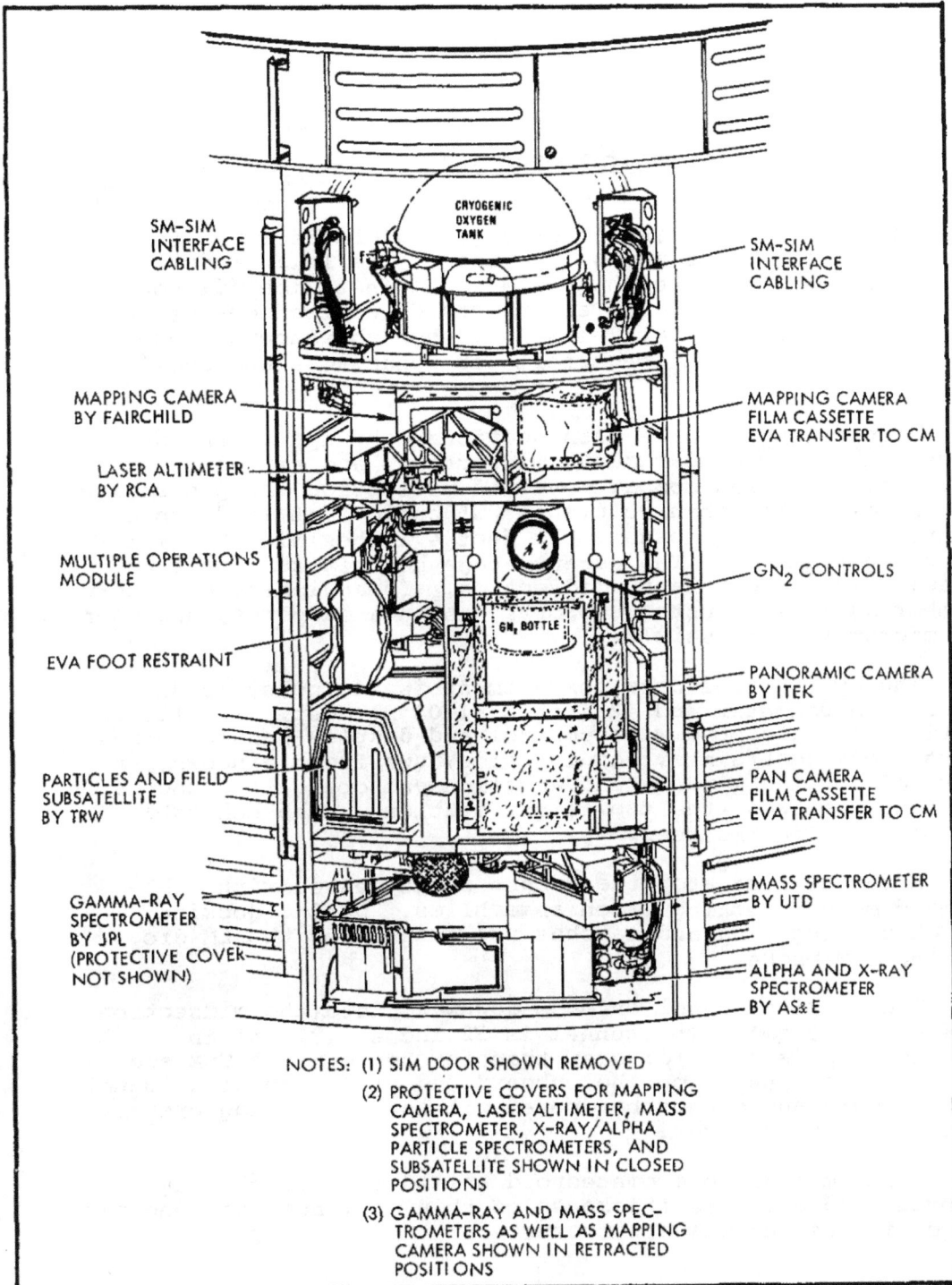

SM-SIM INTERFACE CABLING

MAPPING CAMERA BY FAIRCHILD

LASER ALTIMETER BY RCA

MULTIPLE OPERATIONS MODULE

EVA FOOT RESTRAINT

PARTICLES AND FIELD SUBSATELLITE BY TRW

GAMMA-RAY SPECTROMETER BY JPL (PROTECTIVE COVER NOT SHOWN)

CRYOGENIC OXYGEN TANK

SM-SIM INTERFACE CABLING

MAPPING CAMERA FILM CASSETTE EVA TRANSFER TO CM

GN₂ CONTROLS

PANORAMIC CAMERA BY ITEK

PAN CAMERA FILM CASSETTE EVA TRANSFER TO CM

MASS SPECTROMETER BY UTD

ALPHA AND X-RAY SPECTROMETER BY AS&E

GN₂ BOTTLE

NOTES: (1) SIM DOOR SHOWN REMOVED

(2) PROTECTIVE COVERS FOR MAPPING CAMERA, LASER ALTIMETER, MASS SPECTROMETER, X-RAY/ALPHA PARTICLE SPECTROMETERS, AND SUBSATELLITE SHOWN IN CLOSED POSITIONS

(3) GAMMA-RAY AND MASS SPEC- TROMETERS AS WELL AS MAPPING CAMERA SHOWN IN RETRACTED POSITIONS

Mission SIM Bay Science Equipment Installation

-more-

Lunar Module (LM)

The lunar module is a two-stage vehicle designed for space operations near and on the Moon. The lunar module stands 22 feet 11 inches high and is 31 feet wide (diagonally across landing gear). The ascent and descent stages of the LM operate as a unit until staging, when the ascent stage functions as a single spacecraft for rendezvous and docking with the CM.

Ascent Stage--Three main sections make up the ascent stage: the crew compartment, midsection, and aft equipment bay. Only the crew compartment and midsection are pressurized (4.8 psig). The cabin volume is 235 cubic feet (6.7 cubic meters). The stage measures 12 feet 4 inches high by 14 feet 1 inch in diameter. The ascent stage has six substructural areas: crew compartment, midsection, aft equipment bay, thrust chamber assembly cluster supports, antenna supports, and thermal and micrometeoroid shield.

The cylindrical crew compartment is 92 inches (2.35 meters) in diameter and 42 inches (1.07 m) deep. Two flight stations are equipped with control and display panels, arm-rests, body restraints, landing aids, two front windows, an overhead docking window, and an alignment optical telescope in the center between the two flight stations. The habitable volume is 160 cubic feet.

A tunnel ring atop the ascent stage meshes with the command module docking latch assemblies. During docking, the CM docking ring and latches are aligned by the LM drogue and the CSM probe.

The docking tunnel extends downward into the midsection 16 inches (40 cm). The tunnel is 32 inches (81 cm) in diameter and is used for crew transfer between the CSM and LM. The upper hatch on the inboard end of the docking tunnel opens inward and cannot be opened without equalizing pressure on both hatch surfaces.

A thermal and micrometeoroid shield of multiple layers of Mylar and a single thickness of thin aluminum skin encases the entire ascent stage structure.

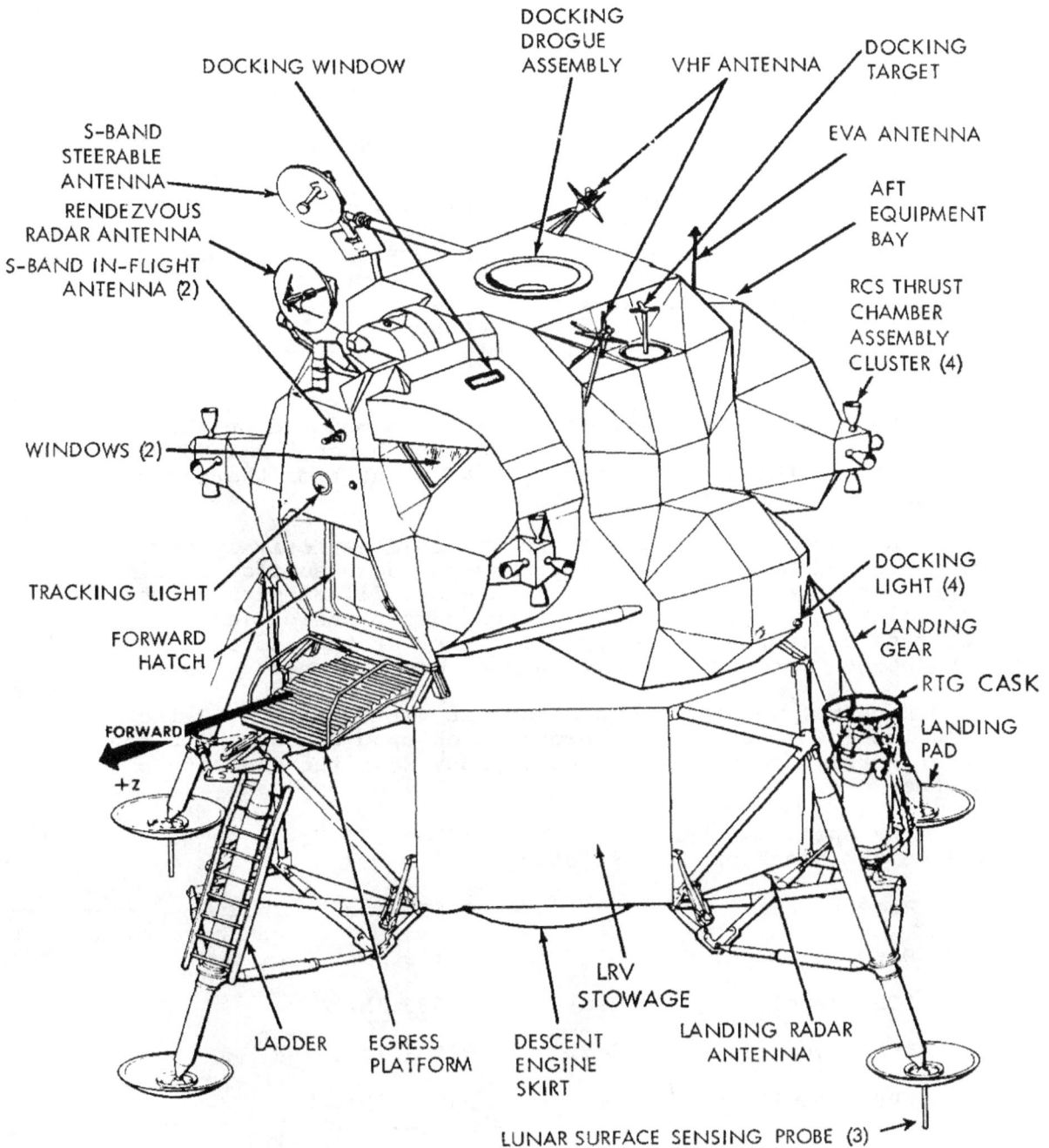

DOCKING WINDOW

DOCKING DROGUE ASSEMBLY

VHF ANTENNA

DOCKING TARGET

S-BAND STEERABLE ANTENNA

RENDEZVOUS RADAR ANTENNA

S-BAND IN-FLIGHT ANTENNA (2)

EVA ANTENNA

AFT EQUIPMENT BAY

RCS THRUST CHAMBER ASSEMBLY CLUSTER (4)

WINDOWS (2)

TRACKING LIGHT

FORWARD HATCH

FORWARD

+Z

DOCKING LIGHT (4)

LANDING GEAR

RTG CASK

LANDING PAD

LADDER

EGRESS PLATFORM

DESCENT ENGINE SKIRT

LRV STOWAGE

LANDING RADAR ANTENNA

LUNAR SURFACE SENSING PROBE (3)

LUNAR MODULE

-more-

Descent Stage--The descent stage center compartment houses the descent engine, and descent propellant tanks are housed in the four bays around the engine. Quadrant II contains ALSEP. The radioisotope thermoelectric generator (RTG) is externally mounted. Quadrant IV contains the MESA. The descent stage measures ten feet seven inches high by 14 feet 1 inch in diameter and is encased in the Mylar and aluminum alloy thermal and micrometeoroid shield. The LRV is stowed in Quadrant I.

The LM egress platform or "porch" is mounted on the forward outrigger just below the forward hatch. A ladder extends down the forward landing gear strut from the porch for crew lunar surface operations.

The landing gear struts are released explosively and are extended by springs. They provide lunar surface landing impact attenuation. The main struts are filled with crushable aluminum honeycomb for absorbing compression loads. Footpads 37 inches (0.95 m) in diameter at the end of each landing gear provide vehicle support on the lunar surface.

Each pad (except forward pad) is fitted with a 68-inch long lunar surface sensing probe which upon contact with the lunar surface signals the crew to shut down the descent engine.

The Apollo LM has a launch weight of 36,230 pounds. The weight breakdown is as follows:

Ascent stage, dry	4,690 lbs.	Includes water and oxygen; no crew
Descent stage, dry	6,179 lbs.	
RCS propellants (loaded)	633 lbs.	
DPS propellants (loaded)	19,508 lbs.	
APS propellants (loaded)	5,220 lbs.	
	36,230 lbs.	

Lunar Module Changes

Although the lunar module exhibits no outward significant change in appearance since Apollo 14, there have been
numerous modifications and changes to the spacecraft in its
evolution from the H mission to the longer-duration J mission
model. Most of the changes involve additional consumables
required for the longer stay on the lunar surface and the
additional propellant required to land the increased payload
on the Moon

Significant differences between LM-8 (Apollo 14) and
LM-10 (Apollo 15) are as follows:

*Fifth battery added to descent stage for total 2075
amp hours. Batteries upgraded from 400 AH each to 415 AH.

*Second descent stage water tank added for total 377
pounds capacity.

*Second descent stage gaseous oxygen (GOX) tank added
for total 85 pounds capacity. Permits six 1410 psi PLSS
recharges. (LM-8: six 900 psi recharges.)

*Addition of system capable of storing 1200 cc/man/day
urine and 100 cc/man/hour PLSS condensate.

*Additional thermal insulation for longer stay time
(67 hours instead of 35 hours on LM-8).

*Additional descent stage payload: Lunar Roving Vehicle
in Quad I previously occupied by erectable S-band antenna and
laser reflector; two pallets in Quad III---one 64.6 pounds
for LRV (holds hand tool carrier) and one 100-pound payload
pallet for laser ranging retro reflector; 600-pound gross
weight capability of enlarged modular equipment stowage
assembly (MESA) in Quad IV, compared to 200-pound capacity
in LM-8; Quad II houses ALSEP.

*Changes to descent engine include quartz-lined engine
chamber instead of silica-lined, and a 10-inch nozzle extension; a 3.36-inch extension to propellant tanks increase
total capacity by 1150 pounds to yield 157 seconds hover
time (LM-8=140 seconds).

MODIFIED LUNAR MODULE
LM-10 AND SUBSEQUENT

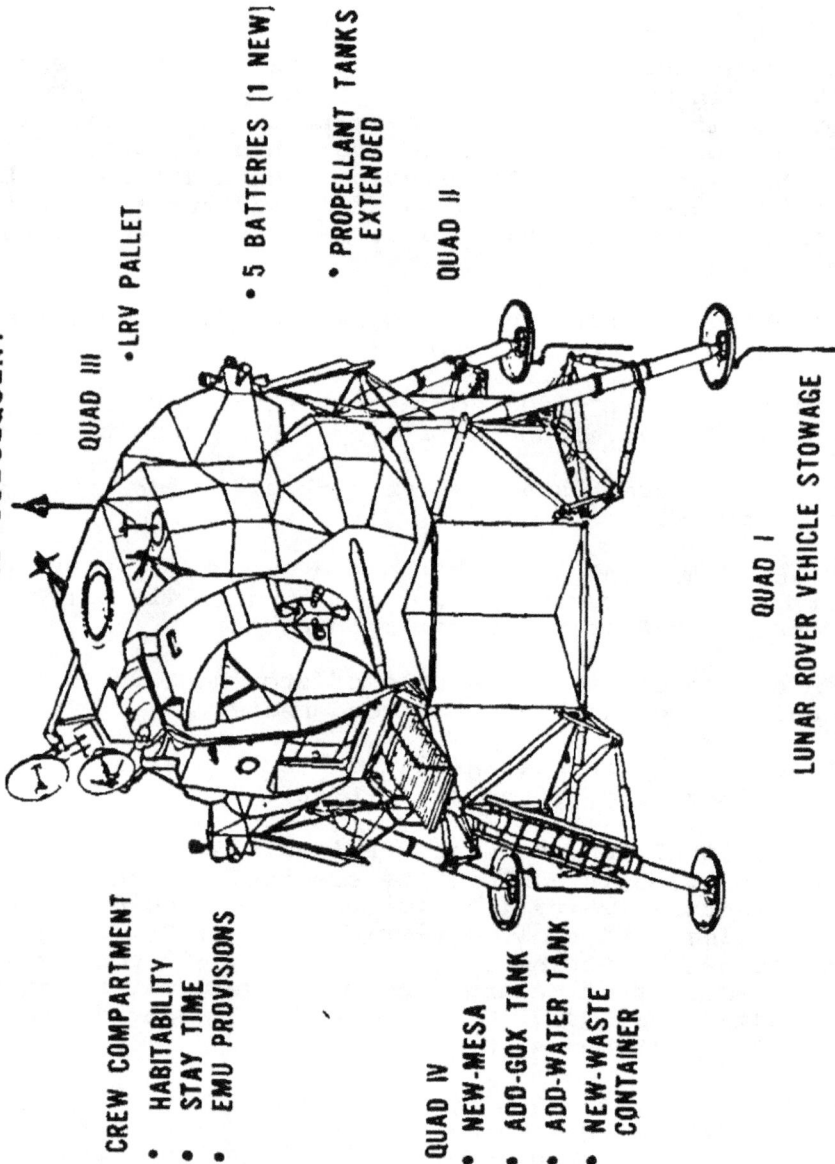

QUAD III
- LRV PALLET

- 5 BATTERIES (1 NEW)
- PROPELLANT TANKS EXTENDED

QUAD II

QUAD I
LUNAR ROVER VEHICLE STOWAGE

CREW COMPARTMENT
- HABITABILITY
- STAY TIME
- EMU PROVISIONS

QUAD IV
- NEW-MESA
- ADD-GOX TANK
- ADD-WATER TANK
- NEW-WASTE CONTAINER

MANNED SPACE FLIGHT NETWORK SUPPORT

NASA's worldwide Manned Space Flight Network (MSFN) will track and provide nearly continuous communications with the Apollo astronauts, their launch vehicle and spacecraft. This network also will continue the communications link between Earth and the Apollo experiments left on the lunar surface, and track the Particles and Fields Subsatellite to be ejected into lunar orbit from the Apollo service module SIM bay.

The MSFN is maintained and operated by the NASA Goddard Space Flight Center, Greenbelt, Md., under the direction of NASA's Office of Tracking and Data Acquisition. Goddard will become the emergency control center if the Houston Mission Control Center is impaired for an extended time.

The MSFN employs 11 ground tracking stations equipped with 30- and 85-foot antennas, an instrumented tracking ship, and four instrumented aircraft. For Apollo 15, the network will be augmented by the 210-foot antenna system at Goldstone, Calif. (a unit of NASA's Deep Space Network), and the 210-foot radio antenna of the National Radio Astronomy Observatory at Parkes, Australia.

NASA Communications Network (NASCOM). The tracking network is linked together by the NASA Communications Network. All information flows to and from MCC Houston and the Apollo spacecraft over this communications system.

The NASCOM consists of more than two million circuit miles, using satellites, submarine cables, land lines, microwave systems, and high frequency radio facilities. NASCOM control center is located at Goddard. Regional communication switching centers are in Madrid; Canberra, Australia; Honolulu; and Guam.

Three Intelsat communications satellites will be used for Apollo 15. One satellite over the Atlantic will link Goddard with Ascension Island and the Vanguard tracking ship. Another Atlantic satellite will provide a direct link between Madrid and Goddard for TV signals received from the spacecraft. The third satellite over the mid-Pacific will link Carnarvon, Canberra, Guam and Hawaii with Goddard through a ground station at Jamesburg, Calif.

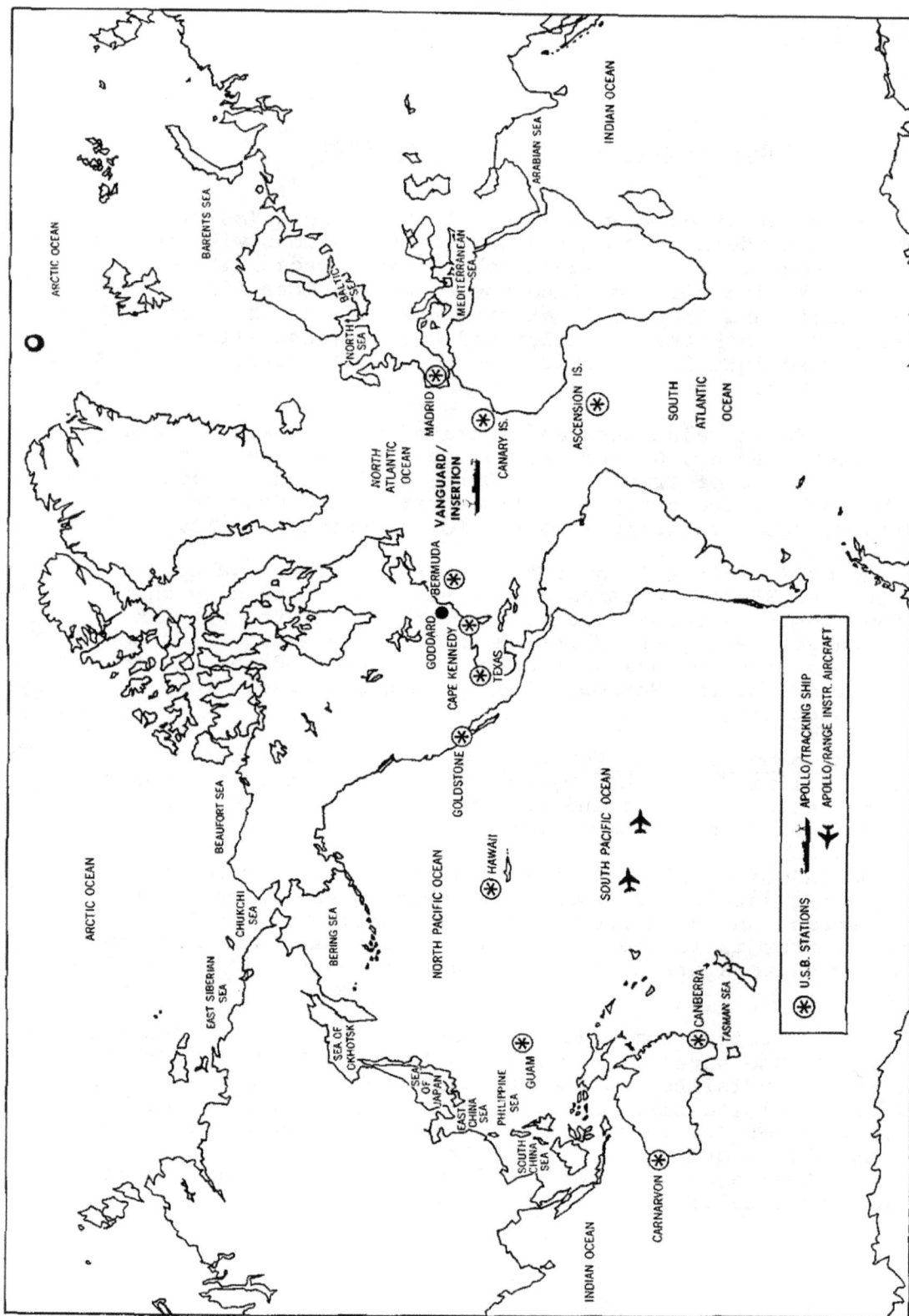

MANNED SPACE FLIGHT TRACKING NETWORK

Mission Operations: Prelaunch tests, liftoff, and Earth orbital flight of the Apollo 15 are supported by the MSFN station at Merritt Island, Fla., four miles from the launch pad.

During the critical period of launch and insertion of the Apollo 15 into Earth orbit, the USNS Vanguard provides tracking, telemetry, and communications functions. This single sea-going station of the MSFN will be stationed about 1,000 miles southeast of Bermuda.

When the Apollo 15 conducts the TLI manuever to leave Earth orbit for the Moon, two Apollo Range Instrumentation Aircraft (ARIA) will record telemetry data from Apollo and relay voice communications between the astronauts and the Mission Control Center at Houston. These aircraft will be airborne between Australia and Hawaii.

Approximately one hour after the spacecraft has been injected into a translunar trajectory, three prime MSFN stations will take over tracking and communicating with Apollo. These stations are equipped with 85-foot antennas.

Each of the prime stations, located at Goldstone, Madrid and Honeysuckle is equipped with dual systems for tracking the command module in lunar orbit and the lunar module in separate flight paths or at rest on the Moon.

For reentry, two ARIA will be deployed to the landing area to relay communications between Apollo and Mission Control at Houston. These aircraft also will provide position information on the Apollo after the blackout phase or reentry has passed.

Television Transmissions: Television from the Apollo spacecraft during the journey to and from the Moon and on the lunar surface will be received by the three prime stations, augmented by the 210-foot antennas at Goldstone and Parkes. The color TV signal must be converted at the MSC Houston. A black and white version of the color signal can be released locally from the stations in Spain and Australia.

TV signals originating from the TV camera stationary on the Moon will be transmitted to the MSFN stations via the lunar module. While the camera is mounted on the LRV, the TV signals will be transmitted directly to the tracking stations as the astronauts tour the Moon.

Once the LRV has been parked near the lunar module, its batteries will have about 80 hours of operating life. This will allow ground controllers to position the camera for viewing the lunar module liftoff, post lift-off geology, and any other desired scenes.

APOLLO PROGRAM COSTS

Apollo manned lunar landing program costs through the first landing, July 1969, totaled $21,349,000,000. These included $6,939,000,000 for spacecraft development and production; $7,940,000,000 for Saturn launch vehicle development and production; $854,000,000 for engine development; $1,137,000,000 for operations support; $541,000,000 for development and operation of the Manned Space Flight Network; $1,810,000,000 for construction of facilities; and $2,128,000,000 for operation of the three Manned Space Flight Centers. At its peak, the program employed about 300,000 people, more than 90 per cent of them in some 20,000 industrial firms and academic organizations. Similarly, more than 90 per cent of the dollars went to industrial contractors, universities and commercial vendors.

Apollo 15 mission costs are estimated at $445,000,000: these include $185,000,000 for the launch vehicle; $65,000,000 for the command/service module; $50,000,000 for the lunar module; $105,000,000 for operations; and $40,000,000 for the science payload.

A list of major prime contractors and subcontractors for Apollo 15 is available in the News Centers at KSC and MSC.

Distribution of Apollo Estimated Program by Geographic Location
Fiscal Year 1971

Based on Prime Contractor Locations
(Amounts in Millions of Dollars)

```
Alabama ------------------------------------ 97
Alaska ------------------------------------- *
Arizona ------------------------------------ 2
Arkansas ----------------------------------- *
California ---------------------------------202
Colorado ----------------------------------- 3
Connecticut -------------------------------- 11
Delaware ----------------------------------- 5
Florida ------------------------------------161
Georgia ------------------------------------ 1
Hawaii ------------------------------------- 1
Idaho -------------------------------------- *
Illinois ----------------------------------- 2
Indiana ------------------------------------ *
```

*Less than one million dollars

-more-

```
Iowa -------------------------------------   *
Kansas -----------------------------------   *
Kentucky ---------------------------------   *
Louisiana --------------------------------  31
Maine ------------------------------------   *
Maryland ---------------------------------   5
Massachusetts ----------------------------  37
Michigan ---------------------------------  35
Minnesota --------------------------------   2
Mississippi ------------------------------  20
Missouri ---------------------------------   2
Montana ----------------------------------   *
Nebraska ---------------------------------   *
Nevada -----------------------------------   *
New Hampshire ----------------------------   *
New Jersey -------------------------------  11
New Mexico -------------------------------   5
New York ---------------------------------  85
North Carolina ---------------------------   *
North Dakota -----------------------------   *
Ohio -------------------------------------   4
Oklahoma ---------------------------------   *
Oregon -----------------------------------   *
Pennsylvania -----------------------------   5
Rhode Island -----------------------------   *
South Carolina ---------------------------   *
South Dakota -----------------------------   *
Tennessee --------------------------------   2
Texas ------------------------------------ 140
Utah -------------------------------------   *
Vermont ----------------------------------   *
Virginia ---------------------------------   4
Washington -------------------------------   6
West Virginia ----------------------------   *
Wisconsin --------------------------------  11
Wyoming ----------------------------------   *
District of Columbia ---------------  21
```

*Less than one million dollars

ENVIRONMENTAL IMPACT OF APOLLO/SATURN V MISSION

Studies of NASA space mission operations have concluded that Apollo does not significantly effect the human environment in the areas of air, water, noise or nuclear radiation.

During the launch of the Apollo/Saturn V space vehicle, products exhausted from Saturn first stage engines in all cases are within an ample margin of safety. At lower altitudes, where toxicity is of concern, the carbon monoxide is oxidized to carbon dioxide upon exposure at its high temperature to the surrounding air. The quantities released are two or more orders of magnitude below the recognized levels for concern in regard to significant modification of the environment. The second and third stage main propulsion systems generate only water and a small amount of hydrogen. Solid propellant ullage and retro rocket products are released and rapidly dispersed in the upper atmosphere at altitudes above 43.5 miles (70 kilometers). This material will effectively never reach sea level and, consequently, poses no toxicity hazard.

Should an abort after launch be necessary, some RP-1 fuel (kerosene) could reach the ocean. However, toxicity of RP-1 is slight and impact on marine life and waterfowl are considered negligible due to its dispersive characteristics. Calculations of dumping an aborted SIC stage into the ocean showed that spreading and evaporating of the fuel occurred in one to four hours.

There are only two times during a nominal Apollo mission when above normal overall sound pressure levels are encountered. These two times are during vehicle boost from the launch pad and the sonic boom experienced when the spacecraft enters the Earth's atmosphere. Sonic boom is not a significant nuisance since it occurs over the mid-Pacific Ocean.

NASA and the DOD have made a comprehensive study of noise levels and other hazards to be encountered for launching vehicles of the Saturn V magnitude. For uncontrolled areas the overall sound pressure levels are well below those which cause damage or discomfort. Saturn launches have had no deleterious effects on wildlife which has actually increased in the NASA-protected areas of Merritt Island.

A source of potential radiation hazard is the fuel capsule of the radioisotope thermoelectric generator (supplied by the AEC) which provides electric power for Apollo lunar surface experiments. The fuel cask is designed so that no contamination can be released during normal operations or as a result of the maximum credible accident.

-more-

PROGRAM MANAGEMENT

The Apollo Program is the responsibility of the Office of Manned Space Flight (OMSF), National Aeronautics and Space Administration, Washington, D.C. Dale D. Myers is Associate Administrator for Manned Space Flight.

NASA Manned Spacecraft Center (MSC), Houston, is responsible for development of the Apollo spacecraft, flight crew training, and flight control. Dr. Robert R. Gilruth is Center Director.

NASA Marshall Space Flight Center (MSFC), Huntsville, Ala., is responsible for development of the Saturn launch vehicles. Dr. Eberhard F. M. Rees is Center Director.

NASA John F. Kennedy Space Center (KSC), Fla., is responsible for Apollo/Saturn launch operations. Dr. Kurt H. Debus is Center Director.

The NASA Office of Tracking and Data Acquisition (OTDA) directs the program of tracking and data flow on Apollo. Gerald M. Truszynski is Associate Administrator for Tracking and Data Acquisition.

NASA Goddard Space Flight Center (GSFC), Greenbelt, Md., manages the Manned Space Flight Network and Communications Network. Dr. John F. Clark is Center Director.

The Department of Defense is supporting NASA during launch, tracking, and recovery operations. The Air Force Eastern Test Range is responsible for range activities during launch and down-range tracking. Recovery operations include the use of recovery ships and Navy and Air Force aircraft.

Apollo/Saturn Officials

NASA Headquarters

Dr. Rocco A. Petrone	Apollo Program Director, OMSF
Chester M. Lee (Capt., USN, Ret.)	Apollo Mission Director, OMSF
John K. Holcomb (Capt., USN, Ret.)	Director of Apollo Operations, OMSF
Lee R. Scherer (Capt., USN, Ret.)	Director of Apollo Lunar Exploration, OMSF

Kennedy Space Center

Miles J. Ross	Deputy Center Director
Walter J. Kapryan	Director of Launch Operations
Raymond L. Clark	Director of Technical Support
Robert C. Hock	Apollo/Skylab Program Manager
Dr. Robert H. Gray	Deputy Director, Launch Operations
Dr. Hans F. Gruene	Director, Launch Vehicle Operations
John J. Williams	Director, Spacecraft Operations
Paul C. Donnelly	Launch Operations Manager
Isom A. Rigell	Deputy Director for Engineering

-more-

Manned Spacecraft Center

Dr. Christopher C. Kraft, Jr.	Deputy Center Director
Col. James A. McDivitt (USAF)	Manager, Apollo Spacecraft Program
Donald K. Slayton	Director, Flight Crew Operations
Sigurd A. Sjoberg	Director, Flight Operations
Milton L. Windler	Flight Director
Gerald D. Griffin	Flight Director
Eugene F. Kranz	Flight Director
Glynn S. Lunney	Flight Director
Dr. Charles A. Berry	Director, Medical Research and Operations

Marshall Space Flight Center

Dr. Eberhard Rees	Director
Dr. William R. Lucas	Deputy Center Director, Technical
R. W. Cook	Deputy Center Director, Management
James T. Shepherd	Director (acting), Program Management
Herman F. Kurtz	Manager (acting), Mission Operations Office
Richard G. Smith	Manager, Saturn Program Office
Matthew W. Urlaub	Manager, S-IC Stage, Saturn Program Office

-more-

Marshall Space Flight Center (cont'd.)

William F. LaHatte	Manager, S-II Stage, Saturn Program Office
Charles H. Meyers	Manager, S-IVB Stage, Saturn Program Office
Frederich Duerr	Manager, Instrument Unit, Saturn Program Office
William D. Brown	Manager, Engine Program Office
S. F. Morea	Manager, LRV Project, Saturn Program Office

Goddard Space Flight Center

Ozro M. Covington	Director, Networks
William P. Varson	Chief, Network Computing & Analysis Division
H. William Wood	Chief, Network Operations Division
Robert Owen	Chief, Network Engineering Division
L. R. Stelter	Chief, NASA Communications Division

Department of Defense

Maj. Gen. David M. Jones (USAF)	DOD Manager for Manned Space Flight Support Operations
Col. Kenneth J. Mask (USAF)	Deputy DOD Manager for Manned Space Flight Support Operations, and Director, DOD Manned Space Flight Support Office

Department of Defense (Cont'd.)

Rear Adm. Thomas B. Hayward (USN)   Commander, Task Force 130,
                                        Pacific Recovery Area

Rear Adm. Roy G. Anderson (USN)   Commander Task Force 140,
                                        Atlantic Recovery Area

Capt. Andrew F. Huff   Commanding Officer, USS
                                        Okinawa, LPH-3 Primary
                                        Recovery Ship

Brig. Gen. Frank K. Everest, Jr.   Commander Aerospace Rescue
   (USAF)   and Recovery Service

-more-

CONVERSION TABLE

	Multiply	By	To Obtain
Distance:	feet	0.3048	meters
	meters	3.281	feet
	kilometers	3281	feet
	kilometers	0.6214	statute miles
	statute miles	1.609	kilometers
	nautical miles	1.852	kilometers
	nautical miles	1.1508	statute miles
	statute miles	0.86898	nautical miles
	statute miles	1760	yards
Velocity:	feet/sec	0.3048	meters/sec
	meters/sec	3.281	feet/sec
	meters/sec	2.237	statute mph
	feet/sec	0.6818	statute miles/hr
	feet/sec	0.5925	nautical miles/hr
	statute miles/hr	1.609	km/hr
	nautical miles/hr (knots)	1.852	km/hr
	km/hr	0.6214	statute miles/hr
Liquid measure, weight:			
	gallons	3.785	liters
	liters	0.2642	gallons
	pounds	0.4536	kilograms
	kilograms	2.205	pounds
Volume:	cubic feet	0.02832	cubic meters
Pressure:	pounds/sq. inch	70.31	grams/sq. cm

-end-